rémont
Point

Bouley
Bay

Rozel Bay

55

Fliquet Bay

+
Trinity

St Catherine's
Bay

+
St Martin

+
St Saviour
+

50

Royal Bay
of
Grouville

T HELIER

+
Grouville

+
St Clement

Green Island

La Rocque
Point

45

Preface

This handbook describes the geology of Jersey, as depicted on the 1:25 000 map published in 1982, and is intended to be read in conjunction with the map.

A geological survey of Jersey was carried out between 1972 and 1977, under a Natural Environment Research Council research contract, by staff of Queen Mary College, University of London, on behalf of the Institute of Geological Sciences (now the British Geological Survey) and the States of Jersey (Island Development Committee). The field work was directed for Queen Mary College by Professor J. F. Kirkaldy until his retirement in 1974, and then by the late Professor W. W. Bishop, supervision being the responsibility of Dr A. C. Bishop and Dr W. J. French. Mr G. Bisson was the District Geologist in charge for the Institute of Geological Sciences, and throughout the work the survey team benefited from the advice and assistance of Dr A. E. Mourant, FRS.

In the handbook, Dr Bishop has written the Introduction (Chapter 1), and the chapters on plutonic igneous rocks, minor igneous intrusions (incorporating information from Mr G. J. Lees of the University of Keele), and economic geology. The chapter dealing with the Rozel Conglomerate has been written by Dr J. T. Renouf, that on the Jersey Shale Formation is by Dr D. G. Helm, and that on the Quaternary deposits is by Dr D. H. Keen. The chapter on geophysical field surveys has been provided by Professor J. C.

Briden, and Dr R. A. Clark of Leeds University, and contributions on palaeomagnetism by Dr B. A. Duff, also of Leeds University, have been included at appropriate places in the text. Mr Bisson has compiled the chapter on volcanic rocks from the unpublished PhD thesis by Dr G. M. Thomas, and the chapter on geological structure mainly from the doctoral thesis by Dr A. D. Squire and the works of Dr Helm and Dr Thomas. Dr N. J. Snelling has advised concerning the geochronology. Contributions from numerous sources, but particularly from Dr Mourant, have been incorporated. It has been necessary to assume that the reader has a basic knowledge of geology, but a glossary of terms that may be unfamiliar has been appended. The handbook has been edited by Mr Bisson.

It gives me great pleasure to thank all the people concerned for their efforts in bringing this work to completion, and Mr R. B. Skinner, Chief Executive Officer of the Island Development Committee, for his sustained interest and support.

F G Larminie, OBE
Director

British Geological Survey
Keyworth
Nottinghamshire NG12 5GG

5 November 1988

Bibliographical reference
BISHOP. A. C. and BISSON, G.
1989. Classical areas of British geology:
Jersey: description of 1:25 000 Channel Islands
Sheet 2. (London: HMSO for British
Geological Survey.)

Authors

A. C. BISHOP, BSc, PhD
Department of Mineralogy, British Museum
(Natural History), Cromwell Road, London
SW7 5BD

G. BISSON, BSc, ARSM
Little Newcombes, Newton St Cyres, Exeter,
Devon EX5 5AW

Contributors

J. C. Briden, MA, PhD
Natural Environment Research Council
Polaris House, North Star Avenue,
Swindon, Wilts SN2 1ET

R. A. Clark, BSc, PhD
Department of Earth Sciences, University of
Leeds, Leeds LS2 9JT

D. G. Helm, BSc, PhD
Department of Earth Sciences, Goldsmiths'
College, University of London, Rachel
McMillan Building, Creek Road, London
SE8 3BU

D. H. Keen, BSc, PhD
Department of Geography, Coventry (Lanchester)
Polytechnic, Priory Street, Coventry CV1 5FB

G. J. Lees, BSc
Department of Geology, University of Keele,
Keele, Staffordshire ST5 5BG

A. E. Mourant, MA, DPhil, DM,
FRCP, FRS
The Dower House, Maison de Haut, Longue-
ville, St Saviour, Jersey, Channel Islands

J. T. Renouf, BSc, PhD
Maison Petit Port, St Brelade, Jersey, Channel
Islands

G. M. Thomas, BSc, PhD
Esso Norge, PO Box 60, N4033 Forus,
Norway

Printed in the United Kingdom for Her
Majesty's Stationery Office

Dd 238943   C20   3/89

HER MAJESTY'S STATIONERY OFFICE

HMSO publications are available from:

HMSO Publications Centre
(Mail and telephone orders)
PO Box 276, London SW8 5DT
Telephone orders 01-873 9090
General enquiries 01-873 0011
Queueing system in operation for both numbers

HMSO Bookshops
49 High Holborn, London WC1V 6HB
  01-873 0011 (Counter service only)
258 Broad Street, Birmingham B1 2HE
  021-643 3740
Southey House, 33 Wine Street, Bristol
  BS1 2BQ   0272-264306
9 Princess Street, Manchester M60 8AS
  061-834 7201
80 Chichester Street, Belfast BT1 4JY
  0232-238451
71 Lothian Road, Edinburgh EH3 9AZ
  (031) 228 4181

HMSO's Accredited Agents
(see Yellow Pages)

And through good booksellers

BRITISH GEOLOGICAL SURVEY

Keyworth, Nottingham NG12 5GG
Plumtree (060 77) 6111

Murchison House, West Mains Road,
Edinburgh EH9 3LA   031-667 1000

London Information Office,
Geological Museum, Exhibition Road,
London SW7 2DE   01-589 4090

The full range of Survey publications is available
through the Sales Desks at Keyworth and
Murchison House, Edinburgh. Selected items
can be bought at the BGS London Information
Office, and orders are accepted here for all
publications. The adjacent Geological Museum
Bookshop stocks the more popular books for sale
over the counter. Most BGS books and reports
are listed in HMSO's Sectional List 45 and can
be bought from HMSO and through HMSO
agents and retailers. Maps are listed in the BGS
Map Catalogue and Ordnance Survey's Trade
Catalogue, and can be bought from Ordnance
Survey Agents as well as from BGS.

The British Geological Survey carries out the geological
survey of Great Britain and Northern Ireland (the latter
as an agency service for the government of Northern
Ireland), and of the surrounding continental shelf, as
well as its basic research projects. It also undertakes
programmes of British technical aid in geology in
developing countries as arranged by the Overseas
Development Administration.

The British Geological Survey is a component body of the
Natural Environment Research Council.

Maps and diagrams in this book use topography
based on Ordnance Survey mapping.

ISBN 0 11 884458 X

Contents

Figures

Table

Notes

Grid references are given in the form [5750 5638] throughout this handbook; they are related to the UTM co-ordinates printed on the 1:25 000 geological map. Numbers preceded by the letter A refer to photographs in the Geological Survey collections.

Numbers preceded by 'Birm' and 'SRR' relate to samples that have been subjected to radiocarbon dating at Birmingham University Geology Department and at the Scottish Universities Research and Reactor Centre respectively.

The isotopic ages given in this work have been recalculated according to the decay constants recommended by the Subcommission of Geochronology at the International Geological Congress in Sydney, Australia, in 1976. As shown in the table below, these constants differ slightly from those of the original cited works, as do the calculated ages.

Decay scheme	Former decay constant	Adopted decay constant
$^{238}U \rightarrow {}^{206}Pb$	$1.537 \times 10^{-10}/a$	$1.55125 \times 10^{-10}/a$
$^{235}U \rightarrow {}^{207}Pb$	$9.722 \times 10^{-10}/a$	$9.8485 \times 10^{-10}/a$
$^{232}Th \rightarrow {}^{208}Pb$	$4.990 \times 10^{-10}/a$	$4.9475 \times 10^{-10}/a$
$^{87}Rb \rightarrow {}^{87}Sr$	$1.47 \times 10^{-11}/a$	$1.42 \times 10^{-11}/a$
	$1.39 \times 10^{-11}/a$	$1.42 \times 10^{-11}/a$
$^{40}K \rightarrow {}^{40}Ar$	$\lambda_c + \lambda_c'\ 0.585 \times 10^{-10}/a$	$0.581 \times 10^{-10}/a$
	$\lambda_\beta 4.72 \times 10^{-10}/a$	$4.962 \times 10^{-10}/a$
	$^{40}K/K = 0.0119$ atomic%	0.01167 atomic%

λ_β = decay by beta emission

λ_c = decay by electron capture

λ_c' = decay by electron capture and decay to ground state

In Chapter 3, Jersey Volcanic Group, the petrographical nomenclature used is based on the work of Ross and Smith, 1961 (see References, p.112).

Introduction

1

The Channel Islands are situated in the Gulf of St Malo and geologically have much in common with the adjacent French mainland of Normandy and Brittany. Their importance to British geologists lies principally in the magnificent coastal sections that expose rocks which have been virtually unaffected by either Caledonian or Hercynian metamorphism, and which, therefore, preserve a largely intact record of local Precambrian and Palaeozoic events. In addition, because the islands lay to the south of the Pleistocene ice-sheets, the superficial deposits present interesting comparisons and contrasts with those of both Britain and France.

This book describes the geology of Jersey (Figure 1). The oldest rocks in the Channel Islands are confined to the Bailiwick of Guernsey, comprising the islands of Guernsey, Sark, Brecqhou, Herm, Jethou, Alderney, Burhou and the Casquets. Metamorphic rocks – principally schists and gneisses – form the southern part of Guernsey, most of Sark and Brecqhou, and the western part of Alderney. Their structures show that they have undergone several phases of deformation and metamorphism during early Proterozoic time at about 2020 million years (Ma) ago (Calvez, 1976).* This ancient basement is not exposed in Jersey but it is present in Normandy and Brittany, particularly near St Brieuc.

Cogné (1959) gave the name Pentevrian to the rocks near St Brieuc that have yielded dates of 1200 to 900 Ma (Leutwein and Sonet, 1965; Leutwein, 1968), and the name has been extended by some authors to include all pre-900 Ma basement rocks within the Armorican massif. Roach and others (1972) have suggested the name Icartian for the 2020 Ma event represented by the Icart orthogneiss of Guernsey.

A succession of low-grade metasedimentary rocks of deep-water continental slope or rise origin and of unknown thickness is widely exposed in Normandy and Brittany, and is

---

* Dr N. J. Snelling comments: Rb:Sr determinations by Adams (1976) led him to suggest that the Icart gneisses were formed at about 2600 Ma ago and underwent a later metamorphism at about 2000 Ma. However, the data were ambiguous and Adams's conclusions were tentative. Subsequent U:Pb determinations by Calvez (1976) clearly indicated an age of 2020 ± 15 Ma for the formation of these gneisses, with no evidence of an Archaean event at about 2600 Ma. (See also Note on p.vii).

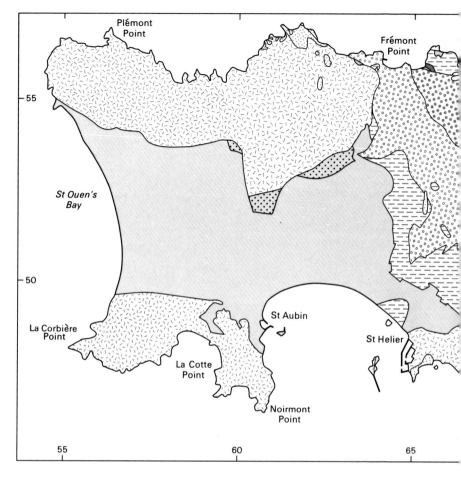

Figure 1 Geological sketch map showing the solid rocks of Jersey

known as the Brioverian (Barrois, 1895). Near St Brieuc
similar metasedimentary rocks rest unconformably on Pentevrian high-grade metamorphic rocks. The oldest rocks in
Jersey – the Jersey Shale Formation – are ascribed to the
Brioverian and were correlated by Graindor (1957) with the
Upper Brioverian of Normandy. There is a raft of Brioverian
metasedimentary rocks in the L'Erée adamellite at Pleinmont
in Guernsey, indicating that these rocks were once more widely distributed in the Channel Islands area than at present. The
Jersey Shale Formation is overlain by a sequence of andesitic
and rhyolitic pyroclastic rocks and lavas that comprise the
Jersey Volcanic Group. Felsic volcanic rocks of broadly
similar age occur at St Germain-le-Gaillard in the Cotentin
and in the Trégor and Erquy areas of Brittany. The Brioverian rocks were folded and metamorphosed to varying ex-

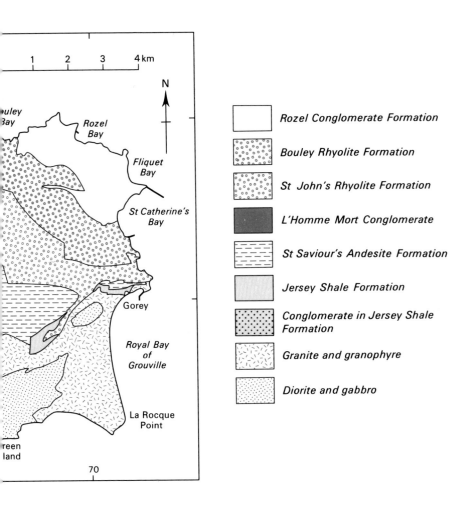

	Rozel Conglomerate Formation
	Bouley Rhyolite Formation
	St John's Rhyolite Formation
	L'Homme Mort Conglomerate
	St Saviour's Andesite Formation
	Jersey Shale Formation
	Conglomerate in Jersey Shale Formation
	Granite and granophyre
	Diorite and gabbro

tents in different places before the deposition of Lower Palaeozoic sediments – the oldest being of Cambrian age – which, in Normandy and Brittany, rest unconformably on them. The name Cadomian is given to the late Precambrian orogeny which deformed the Brioverian rocks.

Igneous rocks occur in each of the Channel Islands, as both major and minor intrusions. Nearly all are related to the Cadomian orogeny, and were intruded in late Precambrian to early Palaeozoic times over the period 675 to 480 Ma (Adams, 1976). Most were emplaced after the Cadomian deformation. As a group, the Jersey granites are somewhat younger than their counterparts in Guernsey, Herm and Alderney, though all have associated gabbroic and dioritic rocks which show many interesting and unusual features. The L'Erée adamellite, exposed on the west coast of Guernsey, differs from the other granites in being foliated and containing large feldspar megacrysts. It is the oldest of the Cadomian granites (645 ±

25 Ma, recalculated from 660 ± 25 Ma of Adams, 1976) and was emplaced before the cessation of Cadomian movements. Other foliated granites occur in Sark and on the islets to the north and south of Jersey (Les Paternosters, Les Dirouilles, Les Ecréhous and Les Minquiers). Isotopic ages show these to have been emplaced at about 617 Ma, before the emplacement of the principal granite masses of the islands.

Other than the inclusion in the granite at L'Erée, rocks corresponding to the Jersey Shale Formation and the overlying volcanic rocks are not exposed in the other islands. The islands comprising the two Bailiwicks are, therefore, geologically complementary, Jersey showing late Precambrian supracrustal sedimentary and volcanic rocks and the effects on them of Cadomian deformation and low-grade metamorphism, in contrast to Guernsey, Sark and Alderney, where the old crystalline basement, predating the Brioverian rocks by about 1000 Ma, is exposed. Cadomian igneous complexes are a common element, and post-Cadomian molasse-type sediments of the Rozel Conglomerate Formation in Jersey are probably broadly coeval with the Alderney Sandstone Formation, which has long been correlated with the earliest Palaeozoic rocks of the Cap de la Hague area of France.

For many years before the present framework of isotopic dates was established, British geologists sought to relate rocks in the Channel Islands to similar formations in south-west England. It is now apparent that such comparisons as can be made with Britain link Brioverian-Cadomian deposits with rocks of broadly similar age exposed, for example, in Anglesey, Shropshire, and near Rosslare in Eire. Granitic rocks of similar age to those in the Channel Islands occur in the Avalon Peninsula of eastern Newfoundland. It seems likely that these areas had similar geological settings in late Precambrian – early Palaeozoic times.

The flat, peneplaned surfaces of the islands bear testimony to much more recent geological events which have shaped the present topography. The loess that occurs on all the main islands is the westerly extension of the European loess, deposited during Pleistocene times by winds blowing across the cold lands to the south of the ice-covered parts of northern Europe. The other superficial deposits reflect both periglacial conditions and episodes of warmer climates and higher sea levels. The numerous raised-beach deposits and evidence of hominid occupation of cave-sites give the islands a significant place in charting the history of Pleistocene and Holocene times.

Jersey Shale Formation

2

The oldest rocks exposed in Jersey belong to the Jersey Shale Formation, formerly known as the Jersey Shales. Both names are unsatisfactory, firstly because the sequence consists of siltstones, sandstones and conglomerates as well as mudrocks, and secondly because the rocks have undergone low-grade regional metamorphism and have been affected by tectonic processes. However, the name Jersey Shale Formation has been adopted to retain continuity with the older nomenclature.

No radiometric dates have been obtained from the Jersey Shale Formation, though it is clearly younger than 1100 – 900 Ma, the minimum age of the Pentevrian at St Brieuc, and older than the oldest Jersey granite, which gave a date of 570 Ma (recalculated from 580 Ma of Adams, 1976). A younger age of 522 Ma (originally 533 Ma), calculated by Duff (1978) for the Jersey Volcanic Group that overlies the Jersey Shale Formation, has been challenged by Bishop and Mourant (1979; see p.58).

On general petrological and stratigraphical grounds Graindor (1957) correlated the Jersey Shale Formation with the Upper Brioverian of Normandy. The discovery of *Sabellarites*-like trace fossils led Squire (1973) to suggest an age of about 750 Ma, but Downie (*in* Bishop and others, 1975) preferred a late Proterozoic (latest Precambrian) Vendian age of between 680 and 570 Ma. However, Bland and others (1987) have shown that the supposed Precambrian fossils described by Squire are actually attributable to modern polychaetous annelids, probably *Polydora sp.*, which made U-form tubes along pre-existing joint planes and veins.

Outcrop distribution and exposure

The main outcrop of the Jersey Shale Formation (Figure 1) occupies a roughly rectangular area extending eastward from St Ouen's Bay to just beyond St Helier, southward to St Aubin's Bay, and northward to La Ville ès Viberts [598 538] and Handois [636 538]. The eastern margin of the outcrop is marked by the disconformable base of the overlying volcanic rocks. The Jersey Shale Formation is bounded to the north by the north-west granite (St Mary's and Mont Mado types; Plate 1) and in the south-west by the south-west (Corbière

Plate 1 Jersey Shale Formation strata (on the left) are in contact with north-west granite (on the right) in a disused quarry at L'Etacq, St Ouen. Notebook for scale. (Photograph by Dr D. G. Helm)

type) granite; at St Helier, the contact (mainly concealed) is with the granophyre of Fort Regent and the south-east (Longueville type) granite.

East of the main outcrop there are two inliers, one centred on Le Bourg [6876 4910] and another extending for about 0.75 km inland from Gorey Harbour on the east coast [702 503 to 712 503], where the Jersey Shale Formation is in contact with the granite of La Rocque type. Smaller outcrops occur at, and to the south of, Frémont Point [6406 5636; 634 559; 639 552], in Giffard Bay [6525 5589], and at Long Echet [6533 5620] and Les Rouaux [6570 5620].

The most extensive exposures of the Jersey Shale Formation occur in St Ouen's Bay*, where a band of intertidal reefs about 1 km wide and 4 km long extends almost continuously from near Le Pinacle [5445 5545] to south-west of L'Ouzière [5646 5182]. After an exposure gap of less than 1 km the formation reappears opposite Le Braye [5651 4996]. Most of the other coastal exposures are good, although less extensive, but inland exposure is limited to isolated quarries and road cuttings.

Lithology

The Jersey Shale Formation, which has an estimated minimum thickness of 2500 m (Helm and Pickering, 1985), mainly consists of fine- to medium-grained sandstones with subordinate conglomerates, siltstones and mudstones, each with variable amounts of diagenetic calcite.

The sandstones range from quartz-wacke to greywacke containing between 10 and 20 per cent of matrix. The framework grains consist of up to 70 per cent quartz, 10 to 15 per cent feldspar (mainly microcline and plagioclase), and about 2 per cent accessory minerals including magnetite, ilmenite and hematite. Carbon flakes up to 10 mm across, of unknown origin, have been recorded by Mourant (1940), Robinson (1960) and Squire (1974), and Dr J. T. Renouf (personal communication) found carbon flakes in greywackes exposed [5835 5198] beside the Val de la Mare Reservoir at the time when the dam was being built.

A varied suite of heavy minerals, including apatite, rutile, zircon, epidote, sphene, almandine, spessartine, glaucophane, sillimanite and staurolite, suggests a diverse provenance of acid and intermediate plutonic igneous rocks, basic subvolcanic rocks, and medium- to high-grade metamorphic rocks; Squire (1974) was unable to detect any significant variation in lateral or vertical distribution of the heavy minerals.

Further evidence that the source area was lithologically mixed comes from pebble to cobble conglomerates; these consist mainly of Jersey Shale Formation intraclasts, but also contain fragments of granite, syenite, andesite, rhyolite, basalt, gneiss and schist. The conglomerates occur at the northern end of St Peter's Valley near La Ville ès Viberts [598 538], between Gargate Mill [605 519] and St Anastase [608 524], and around Carrefour Selous [6232 5326] and Quetivel Mill [6317 5323], but exposure is poor.

*Owing to a printing error on the 1:25 000 geological map, an area [561 516] south-west of L'Ouzière has been coloured as dolerite: except for two narrow dolerite dykes it should be shown as Jersey Shale Formation.

Metamorphism

The sediments have been regionally metamorphosed to low greenschist facies, with the conversion of the clay matrix to chlorite. In addition, wherever the formation has been intruded by granite it has undergone contact metamorphism. At the northern end of St Ouen's Bay, a zone of chlorite spotting extends from the La Bouque Fault [545 545] for 500 m north towards the north-west granite, where it passes into a zone of cordierite-biotite-hornfels 5 to 10 m wide. Similar contact aureoles occur elsewhere; for example, at the southern end of St Ouen's Bay, in Gorey Harbour [713 503], where greywacke has been altered to hornblende-hornfels, and at the northern end of St Peter's Valley, where the metasedimentary clasts and originally muddy matrix of the conglomerates show extensive chlorite spotting and local growth of biotite.

Sedimentology

Squire (1974) concluded that the Jersey Shale Formation represented a 'eugeosynclinal submarine fan' (a deep-water, fan-shaped 'delta') formed in a sedimentary basin associated with a subduction zone. Helm and Pickering (1985) supported Squire's general inference, and in addition their work enabled them to recognise several distinct depositional environments within the fan. They described six facies and

Figure 2 Sketch map showing the distribution of facies associations in the Jersey Shale Formation. Redrawn from Helm and Pickering, 1985, fig.15

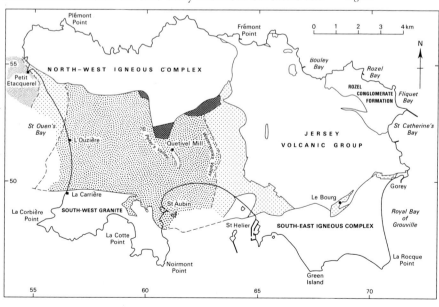

I   II   III   III and/or IV   IV

grouped them into four associations (Figure 2), each of which characterised a particular depositional environment. Association I crops out at three places on the southern margin of the north-west granite. It consists mainly of disorganised clast-supported pebble to cobble conglomerates, which suggest deposition from flows in which the clasts were partly supported by the cohesive strength of the matrix of mud and water, and partly by collisions between each other. No direction of palaeocurrent flow has been determined for this association.

The graded, medium- to fine-grained sandstones and rare cross-bedded, medium-grained, lenticular sandstones of Association II form a lenticular body exposed only at isolated localities near the southern end of St Peter's Valley [610 516]. This association is characterised by a high ratio of sand to shale, amalgamation and splitting of beds, and cross-bedding, together suggesting deposition in channels that cut or grade laterally into rocks of Association III. Limited palaeocurrent data suggest that flow was towards the east or south-east.

Association III, best seen on the foreshore and intertidal zone west of Grand Etacquerel [5470 5480] (Plate 15), is composed mainly of ripple-laminated, very fine-grained sandstones, and graded medium- to fine-grained sandstones (Plate 2). These show the sequence of internal structures typical of deposition from relatively high-density turbidity currents (Bouma, 1962), and also display the effects of soft-sediment, post-depositional liquefaction and flow in the form of distorted cross-laminae and cross-cutting sedimentary 'dykes'. A local facies variant, consisting of granule-grade sandstone with both normal and inverse grading, occurs by the slipway north-north-east of Petit Etacquerel [5480 5470]. Palaeocurrent observations from current ripples indicate flow towards the north-north-west, though some imply flow in the opposite direction. Flute casts (asymmetrical scours) on the bases of sandstone units consistently signify flow towards the north.

Association IV is the most important of the associations in terms of area and volume. It consists of laminated mudstone, ripple-laminated very fine-grained sandstone, graded medium- to fine-grained sandstone, and cross-stratified medium-grained sandstone. The best exposures are in the central and northern parts of St Ouen's Bay. Characteristically, the thickness and coarseness of sandstone beds in the sequences comprising Association IV increase upwards and are accompanied by an increase in the ratio of sand to shale; for example, on the foreshore opposite La Saline [555 540] and at Grand Etacquerel [547 547]. Similar sequences, which also display evidence of folding and sliding in soft sediments, occur on the west side of St Aubin's Bay [607

484]. Palaeocurrent data suggest that flow was generally towards the north.

Environment of deposition

The sedimentary structures and facies suggest that the Jersey Shale Formation accumulated in an aqueous environment below wave-base. The facies associations are interpreted by analogy with the submarine fan model (Mutti and Ricci Lucchi, 1972; Normark, 1978; Walker, 1978, 1984).

Most ancient and modern submarine fans display a number of distinguishing morphological features directly related to their mode of formation. In the classic submarine fan model it is envisaged that a delta-shaped pile of sediment, usually supplied from a single submarine canyon, is banked against the continental slope. As additional sediment is supplied, the fan spreads over and merges with the basin plain.

Plate 2 Intertidal reefs near Petit Etacquerel show Jersey Shale Formation medium to thick amalgamated sandstones with thinly bedded sandstones (bottom left). A chaotic bed occurs just above the notebook. The cliffs in the background consist of north-west granite.
(Photograph by Dr D. G. Helm)

Submarine fans may be divided into: an inner (or upper) fan, with a single deep channel or canyon; a mid-fan, with migrating depositional channels that become shallower away from the inner fan; and a topographically relatively smooth outer (or lower), non-channelled fan, containing mounds of sediment (lobes) deposited at the mouths of channels, which grades into a more or less featureless basin plain. The margins or fringes of the major divisions of the submarine fan are gradational. Overall, radially outwards from the main feeder channel, there is a decrease in sandstone thickness, a general reduction in the amount of channelling, an increase in the ratio of muddy to sandy sediment, and a general reduction in grain-size.

Associations I and II are considered to have been deposited in the axial parts of submarine channels, the coarser grain-size of Association I suggesting that it is nearer to its source of supply than is Association II. Association I appears to have accumulated in a canyon or canyons in the inner fan, whereas Association II seems to have affinities with typical mid-fan deposits (e.g. Pickering, 1983). Association III has the characteristics of sedimentation in shallow ephemeral channels with some interchannel and/or fan-fringe deposits, i.e. in the lower part of the mid-fan to outer-fan environment. Association IV has many of the attributes of a typical outer fan (Mutti and Ricci Lucchi, 1972; Pickering, 1981), characterised by lobe, lobe-fringe and fan-fringe deposits.

The regional setting

Helm and Pickering (1985) found that the Jersey Shale Formation youngs overall to the east. From the general distribution of the facies associations (Figure 2) and the limited palaeocurrent data, they concluded that the submarine fan was constructed from mainly northerly-directed. sediment gravity flows.

Thomas (1977) recorded that the Jersey volcanic rocks (Chapter 3) display a calc-alkaline compositional trend (typical of island-arc volcanism); that volcanic activity of comparable age was widespread in south Britain and north Brittany; and that ancient metamorphic rocks form the basement beneath thick sedimentary successions of Longmyndian and Brioverian age in Anglesey and north-west France respectively. These considerations led him to suggest, in the light of plate-tectonic reconstructions by Dewey (1969) and Mitchell and Reading (1971), that the Jersey volcanic rocks (and by inference the Jersey Shale Formation) accumulated on continental basement about 500 km south of a subduction zone dipping to the south associated with an Andean-type continental plate margin.

Helm and Pickering (1985) concluded that the Jersey Shale Formation probably represents submarine-fan deposition on a continental margin. The shape and size of the deepwater sedimentary fan complex was difficult to assess because of the isolation and small size of Jersey. However, the sedimentology of the Jersey Shale Formation, its relationship to the overlying volcanic rocks, and the inferred regional setting, are all consistent with deposition in a basin perched on a continental slope adjacent to a volcanic arc above a subduction zone.

Jersey Volcanic Group   3

The deposition of the Jersey Shale Formation was followed by the eruption and accumulation of a thick succession of volcanic and volcaniclastic rocks. The earliest volcanic rocks were andesitic in composition, and these gave way to rhyolitic rocks, porphyritic at first and then aphyric. The following sequence, with estimated thicknesses, has been recognised (Thomas, 1977), broadly following Mourant (1933):

Bouley Rhyolite Formation (aphyric)	430 m
St John's Rhyolite Formation (porphyritic)	950 m
St Saviour's Andesite Formation	850 m

The boundary between the Jersey Shale Formation and the St Saviour's Andesite is disconformable overall, but seems to be conformable locally (see p.15–16, 20). Thus the volcanic rocks are probably only a little younger than the greywackes and may be of late Precambrian age. However, Duff (1978) has determined the age of porphyritic amygdaloidal andesites from the West Mount Quarry, just west of St Helier, as 533 ± 16 Ma (522 ± 16 Ma when recalculated), and has concluded that these rocks are Cambrian in age, although this opinion has been contested by Bishop and Mourant (1979; see p.58).

The main outcrop of volcanic rocks extends from St John's Bay in the north, to Anne Port in the east. These rocks have been folded into a broad synclinorium (the Trinity Syncline; Figure 18), plunging NNE, and they are separated by the Frémont Fault from a smaller outcrop of andesites south of Belle Hougue Point. The andesites west of St Helier form a syncline (the St Helier Syncline) plunging SSW.

The andesitic lavas and pyroclastic rocks are typical of the deposits laid down on the flanks of volcanoes, and it is likely that there were vents to the north-east and south of the existing outcrops. However, because of their variable nature, it is possible to make only broad correlations between the sequences in different areas.

There is a marked angular discordance between the andesites and the succeeding rhyolites. This partly reflects differing modes of eruptive activity, as the andesitic volcanism formed cones whereas the rhyolitic ignimbrite flows tended to fill topographic lows; however, the presence of interformational laharic mudstones and conglomerates suggests that there was a short period of erosional activity between deposition of the formations.

Apart from a few local andesitic lavas and pyroclastic rocks, the St John's Rhyolite Formation comprises five massive ignimbrite cooling units, each containing several individual flows. There is a possible fissure feeder for the earliest member, the Jeffrey's Leap Ignimbrite, in Queen's Valley, but the source of the other ignimbrites is uncertain.

Flows of rhyolite marked the start of deposition of the Bouley Rhyolite Formation. Thin andesites occur among the rhyolites, and there is evidence of erosion at some levels in the sequence. In the eastern coastal section the rhyolites are succeeded by fairly thin aphyric ignimbrite flows with intercalated air-fall tuffs, all folded about E – W axes. A late porphyritic ignimbrite, the St Catherine's Ignimbrite, overlies the sequence with angular discordance. Ignimbrites around Bouley Bay are similar to the aphyric material from the east coast, but folded about N – S as well as E – W axes. The whole volcanic succession is unconformably overlain by the Rozel Conglomerate.

An intrusive contact between the volcanic rocks and the north-west granite is well exposed at Côtil Point [631 563]. A thermal aureole about 300 m wide features a pronounced N – S vertical foliation, granoblastic textures, and the growth of biotite porphyroblasts. Metasomatic andradite, diopside, and epidote are locally present in veins and patches, and the volcanic rocks close to the granite have been metasomatically enriched in potassium.

St Saviour's Andesite Formation

The St Saviour's Andesite Formation consists of subaerially deposited lavas, tuffs, and agglomerates of andesitic and basaltic composition. The andesites have a keratophyric mineralogy which is thought to be the result of propylitisation rather than regional metamorphism. All the lavas contain numerous plagioclase phenocrysts which, except in one basalt, have been altered to albite. The basalts all contained original olivine phenocrysts and most of the andesites original pyroxene, though in a few cases amphibole occurred instead.

Belle Hougue area   The successions found in the Belle Hougue area (Figure 3) are shown diagrammatically in Figure 4. The breccio-conglomerate at the base is a variable deposit, up to 25 m thick, which separates the Jersey Shale Formation from the volcanic sequence. The earliest beds of the breccio-conglomerate contain fragments of Jersey Shale

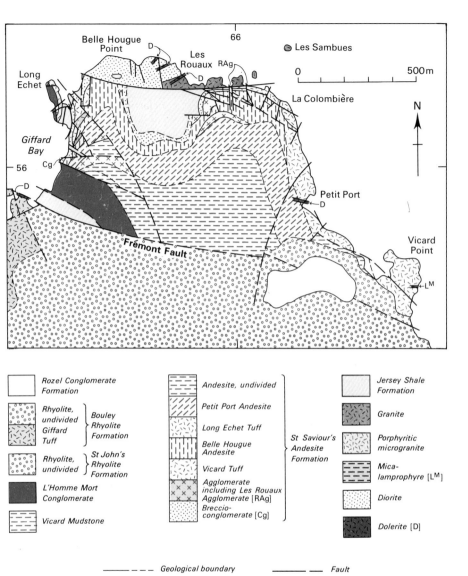

Rozel Conglomerate Formation

Rhyolite, undivided
Giffard Tuff } Bouley Rhyolite Formation

Rhyolite, undivided } St John's Rhyolite Formation

L'Homme Mort Conglomerate

Vicard Mudstone

Andesite, undivided

Petit Port Andesite

Long Echet Tuff

Belle Hougue Andesite

Vicard Tuff

Agglomerate including Les Rouaux Agglomerate [RAg]

Breccio-conglomerate [Cg]

St Saviour's Andesite Formation

Jersey Shale Formation

Granite

Porphyritic microgranite

Mica-lamprophyre [LM]

Diorite

Dolerite [D]

——— – – – Geological boundary   ——— — Fault

Figure 3 Sketch map of the geology south of Belle Hougue Point. Based on Thomas, 1977, fig.2.2

Formation only, but later beds have fragments of andesite and pebbles and grains of quartz in increasing amounts, producing a grit, in places conglomeratic, with thin mudstone lenses. The upper part at least of the breccio-conglomerate has the appearance of a braided stream deposit. A lens of grey-green lava 10 m long and 1 m thick is present [6540 5630] within the grits; this lava may be an altered basalt and was probably deposited in water in a river bed or shallow lake. Dips and strikes in the breccio-conglomerate nowhere show distinct divergence from those in the Jersey Shale Formation, and it is likely that the volcanic succession followed

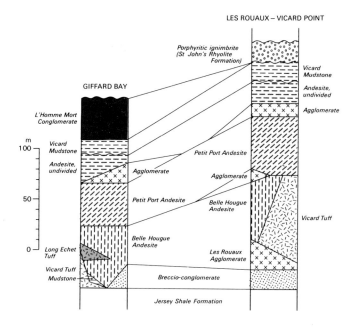

Figure 4
Generalised
vertical sections at
Giffard Bay and
between Les
Rouaux and
Vicard Point.
Based on Thomas,
1977

conformably on the Brioverian sedimentary rocks, though
Helm (1984) has argued for a discontinuity on the grounds
that the presence of a thick conglomerate at the base of the
volcanic succession, which also appears to overstep some
units of the Jersey Shale Formation, indicates that there was
a considerable time interval between deposition of the two
sequences.

The Les Rouaux Agglomerate is up to 25 m thick. In its
type area it has a sharp base. Angular fragments of shale,
andesite, porphyry and dark green pumice are set in a
purplish grey matrix, the pumice being confined to the
earliest beds. The matrix has undergone felsitic devitrification. Lenses of green tuff up to 0.15 m thick occur towards
the top of the agglomerate. The top is generally distinct, but
in places it merges with the overlying Vicard Tuff, the grey-
purple matrix becoming bright green with the disappearance
of lithic fragments.

The Vicard Tuff, consisting of grey-green to bright green
and pinkish grey fine-grained tuffs, overlies the breccio-
conglomerate on the east side of Giffard Bay and the Les
Rouaux Agglomerate farther east. It is up to 80 m thick, and
at its top in the Les Rouaux area it is interlayered with the
Belle Hougue Andesite. Bedding is not usually apparent, but
west of Vicard Point the tuff displays crude layering and contains many bombs; these are up to 40 mm long, are usually
aligned parallel to the bedding, and have disrupted the layering on their undersides only.

The Belle Hougue Andesite is exposed on the east side of Giffard Bay, where its base rests on successively lower horizons towards the south, first on the Vicard Tuff and finally on the Jersey Shale Formation; this suggests flow down an eroded palaeoslope. At Les Rouaux the Belle Hougue Andesite overlies the Vicard Tuff; in this area it has a maximum thickness of 60 m, but it thins south of La Colombière, where it is overstepped by the Petit Port Andesite (see below). The Belle Hougue Andesite is normally purple, but it is grey-green in places where the lava was contaminated with Vicard Tuff. The Long Echet Tuff (see below) separates a basal flow 1 m thick from the rest of the andesite, and up to 5.5 m of tuffs of Vicard type within the andesite at Les Rouaux also indicate the presence of at least two flows.

The Long Echet Tuff occurs above the base of the Belle Hougue Andesite on the east side of Giffard Bay. It is up to 15 m thick and consists of megacrysts up to 2 mm in diameter of corroded plagioclase, perthite, and bipyramidal quartz, set in a felsitic matrix. Fiamme, which resemble strands of white cotton in hand specimen, are up to 15 mm long and have been devitrified to granular quartz. Near its base it contains fragments of shale and Vicard Tuff. The Long Echet Tuff, which is markedly more silicic than the surrounding tuffs and lavas, is ignimbritic.

The dark grey-green Petit Port Andesite is up to about 55 m thick at Petit Port [6625 5590], where it rests on Vicard Tuff. Farther west it overlies Belle Hougue Andesite, from which it is locally separated [6595 5625] by a lens of agglomerate up to 12 m thick. It is succeeded by agglomerate south of Belle Hougue Point, and by tuff west of Petit Port; both are overlain by undivided andesite to the north of Egypt, where exposure is poor.

St Helier Syncline   The St Saviour's Andesite Formation in the St Helier Syncline (Figure 5) comprises the following members in downward order:

Bathing Pool Andesites (Upper, Middle and Lower)
Bathing Pool Agglomerate
West Mount Andesite
West Mount Tuff
St John's Road Agglomerate
St John's Road Andesite

The St John's Road Andesite is exposed along Old St John's Road [6475 4950] as a highly weathered limonite-stained lava. White feldspar laths and almost acicular chlorite patches are set in a greyish green groundmass. The flow is probably 30 to 50 m thick. The contact with the overlying St John's Road Agglomerate is not exposed, but 80 m of agglomerate can be examined at the western end of West Park [642 493] and 10 m at the northern end of West Mount Quarry. At the latter locality angular to subrounded andesite fragments up to 1 m across are set in a grey-green fine-grained felsitic matrix. In West Park macroscopic lithic fragments do not exceed 40 mm in diameter, and include shale and pink fine-grained quartz-porphyry as well as andesite.

The West Mount Tuff is pale green and fine grained, and is from 10 to 26 m thick. Exposures in West Park [6435 4920] indicate that the junction with the underlying agglomerate is irregular; the change from agglomerate to tuff is abrupt, and the tuff is therefore thought to have been produced by a separate eruptive event. In West Mount Quarry

Figure 5
Geological sketch map showing the St Helier Syncline. Redrawn from Thomas, 1977, fig.3.1

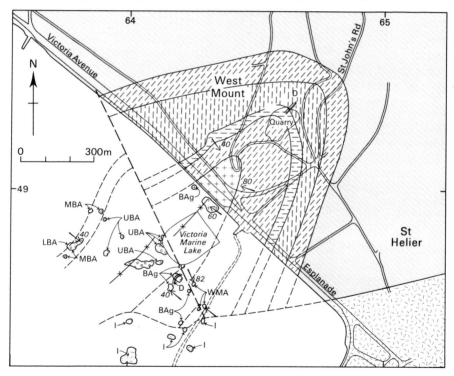

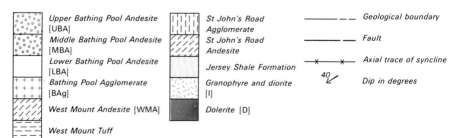

Upper Bathing Pool Andesite [UBA]	St John's Road Agglomerate	Geological boundary
Middle Bathing Pool Andesite [MBA]	St John's Road Andesite	Fault
Lower Bathing Pool Andesite [LBA]	Jersey Shale Formation	Axial trace of syncline
Bathing Pool Agglomerate [BAg]	Granophyre and diorite [I]	Dip in degrees
West Mount Andesite [WMA]	Dolerite [D]	
West Mount Tuff		

the tuff is intimately mixed with the lowest 15 m of the overlying West Mount Andesite, in a manner which suggests that it was originally an unconsolidated ash. The sequence appears to young to the south.

The West Mount Andesite varies in thickness from 36 m in West Park to 100 m in West Mount Quarry. In the quarry amygdaloidal porphyritic lava gives way southwards to phenocryst-poor andesite without amygdales, but the lithology reverts to a highly porphyritic and amygdaloidal type farther south [6452 4918]; therefore there are probably at least three separate lava flows of variable lateral extent within the member. Partly altered albite phenocrysts pick out a poor but recognisable fluxion banding.

The Bathing Pool Agglomerate varies laterally. Around West Park Pavilion greenish purple fine-grained tuff rests on the cracked surface of the West Mount Andesite. On the nearby beach, reefs to north and south of the Victoria Marine Lake (bathing pool) display poorly graded units of tuff and agglomerate from a few centimetres to several metres in thickness. Bedding can rarely be traced for more than a few metres and no aqueous sedimentary textures are present, indicating that it is likely to be an air-fall deposit. The consistent similarity of all the larger andesite fragments suggests that the agglomerate was formed by the explosive disruption of a single homogeneous volcanic plug. Shale and porphyry fragments were probably derived from the walls of the volcanic conduit.

Three separate lava flows and an inter-flow sedimentary horizon have been distinguished in the Bathing Pool Andesites. The Lower Bathing Pool Andesite has corroded platy albite phenocrysts up to 10 mm long, set in a grey fine-grained groundmass, commonly in clusters and showing no particular orientation. Amygdales up to 20 mm long are abundant, and are orientated parallel to the surface of the flow. In a reef [6378 4879] west of the bathing pool the top 7 m of the Lower Bathing Pool Andesite are overlain by olive-brown mudstone up to 0.3 m thick and by the basal 25 m of the Middle Bathing Pool Andesite (Plate 3). Blocks from the underlying andesite occur near the base of the Middle Bathing Pool Andesite, and fragments of the mudstone are distributed throughout its exposed thickness.

The Middle Bathing Pool Andesite is highly porphyritic and fluxion-banded. The rock is readily identified by the abundant greenish white platy feldspar laths which are set parallel to the banding in a greyish green fine-grained groundmass. A few amygdales up to 3 mm across are lined with epidote and have penninite centres.

The Upper Bathing Pool Andesite, the youngest member of the local andesite sequence, is probably about 100 m thick. It forms the core of the St Helier Syncline. On the western limb of the fold it follows the Middle Bathing Pool Andesite, but on the eastern limb the two lowest members of the Bathing Pool Andesites have been overstepped by the youngest lava, which rests directly on tuffs at the top of the Bathing Pool Agglomerate. The base can be examined on a reef [6418 4869] south of the bathing pool. Fluxion banding in the Upper Bathing Pool Andesite is picked out by pinkish white albitised plagioclase laths.

Trinity Syncline The largest outcrop of andesitic rocks in Jersey is situated inland, around the main synclinorium. The boundary between the Jersey Shale Formation and the andesites is exposed at several places. At West Hill [6435 5052] highly weathered exposures show a vertical junction between shales and tuffs; an agglomerate layer 0.2 m thick lies within the shales, and the andesites here are considered to be conformable upon the shales. Faulted junctions have been seen at Chestnut Farm [6465 5000], Clos de Paradis [653 498] and Wellington Road [6612 4925].

At Côtil Point (Figure 7) andesite has been intruded by granite (p.52) and is overlain by ignimbrite. Just north-east of the site of the former Mont Mado Quarry, outcrops of andesite are terminated to the north by a pre-granite E–W sinistral wrench fault. These andesites have been affected by thermal metamorphism; they contain pink alkali feldspars rather than plagioclase, but display no megascopic foliation.

Plate 3 Lower Bathing Pool Andesite (below) and highly porphyritic Middle Bathing Pool Andesite (above) are separated by a thin mudstone bed in a reef west of the Victoria Marine Lake. Limpets show the scale. (A13718)

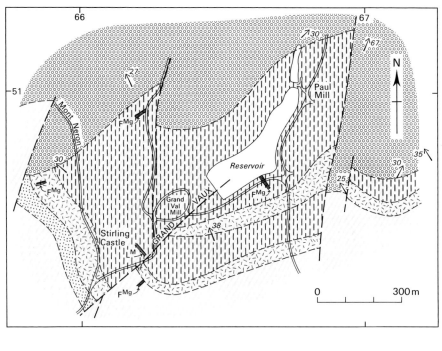

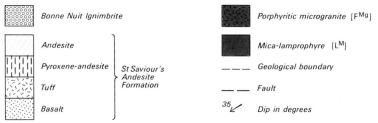

| | Bonne Nuit Ignimbrite | | | | Porphyritic microgranite [F^Mg] |
| | | | | | |

Bonne Nuit Ignimbrite

Andesite ⎫
Pyroxene-andesite ⎬ St Saviour's Andesite Formation
Tuff ⎪
Basalt ⎭

Porphyritic microgranite [F^Mg]

Mica-lamprophyre [L^M]

— — — Geological boundary

— — Fault

35⟋ Dip in degrees

Figure 6 Sketch map of the geology around Grands Vaux Reservoir. Based on Thomas, 1977, fig.4.9

Between Handois Reservoir and Mont à l'Abbé [650 503] the andesites consist of albite phenocrysts in a dark grey-green fine-grained felsitic groundmass rich in plagioclase microlites and commonly fluxion-banded. Mafic phenocrysts have been replaced by chlorite, quartz, epidote and iron ore, but recognisable pseudomorphs after pyroxene occur. Apatite needles are scattered throughout the rock.

In the quarry [6536 5037] behind Homestead Cottage, in Vallée des Vaux, two lava flows can be recognised. Dark green blocky andesite at the north end is distinct from the light green highly amygdaloidal andesite, which contains pyroxene phenocrysts and fewer albite laths, in the rest of the quarry. The distribution of amygdales and rubbly texture in these lava flows suggests that the sequence may be inverted. Tuffs resembling the Vicard Tuff are exposed elsewhere in Vallée des Vaux. Two exposures [6534 5053] north of

Homestead Cottage are in altered basalt, with white platy labradorite phenocrysts and pseudomorphs after poikilitic olivine crystals set in a purple-green fine-grained groundmass. In Grands Vaux valley (Figure 6) grey-green keratophyric pyroxene-andesite is interlayered with tuff and agglomerate, and near Stirling Castle [659 506] basalt similar to that noted in Vallée des Vaux is exposed.

The Trinity Hill Andesite, greenish purple andesite rich in platy feldspar laths, is exposed on New Trinity Hill. Dips on the fluxion banding indicate that the flow is folded into a WNW–ESE-trending syncline, the axis of which lies near the junction with Old Trinity Hill.

Around St Saviour's Hill the succession recognised north of the faulted boundary with the Jersey Shale Formation is, in downward order:

Grey-green pyroxene-andesite with tuff horizons	95 m
Agglomerate	7 m
Purple amygdaloidal andesite	27 m
Grey andesite	16 m

Grey andesite around St Mark's School [6581 4956] has sericitised albite laths set in a fluxion-banded groundmass of plagioclase microlites, interstitial feldspar and finely divided magnetite. The lava is rich in amygdales towards its top, and is overlain by purple Trinity Hill Andesite with a sharp irregular contact. Agglomerate was seen at only one place [6587 4948], just above the purple andesite, and presumably is not extensive. Similarly, an old quarry [6602 4968] in the grounds of Government House provides the only exposure of tuff in the vicinity, the rock being silvery grey, fine grained and compact.

In Swiss Valley the succession north of the granite contact [671 490] consists of tuff, overlain by pyroxene-andesite with two further tuff horizons; agglomerate overlies the uppermost tuff. The andesites resemble those found in Grands Vaux and at St Saviour. A conglomerate up to 7 m thick forms an outcrop just east of Beau Désert [6718 4960]; pebbles of pyroxene-andesite in the conglomerate are of local type and up to 60 mm in diameter, set in a matrix of fine andesite debris. This conglomerate is presumably the product of inter-flow erosion, possibly along a stream bed or gulley. The sporadic occurrence of tourmaline (schorlite) in volcanic rocks between St Saviour's Parish Hall [663 498] and south-east of Francheville [6850 4949] has been recorded (Mourant, 1933; Thomas, 1977); this mineral is assumed to have resulted from pneumatolysis, though the origin of the fluids is uncertain.

An inlier of andesitic rocks around Blanche Pierre [667 524], west of Le Grès, has the form of an eroded dome (D_1; see p.72); the rocks comprise pyroxene-andesite, agglomeratic tuff and tuffite, all surrounded by ignimbrite. The tuffite is a grey-green, compact, fine-grained rock with graded units up to 40 mm thick; quartz grains are present near the base but do not persist to higher levels. The rock appears to have been deposited in quiet lacustrine conditions, and field relations favour association with the andesites rather than the ignimbrites.

A triangular area of purple amygdaloidal lava [699 531] south of Rozel Manor was termed andesite by Mourant (1933), but was identified as altered basalt by Thomas (1977). This inlier provides evidence for the presence of a major fault beneath the Rozel Conglomerate (see p.77).

St John's Rhyolite Formation

The St John's Rhyolite Formation follows the St Saviour's Andesite unconformably, but the discontinuity between the two may reflect overstep rather than a genuine break in the eruptive events. Interformational sediments include the Vicard Mudstone and the L'Homme Mort Conglomerate.

The name Vicard Mudstone is given to a purple and green mudstone found at Vicard Bay (where it is 60 m thick, inverted, and dips steeply to the north) and also above Giffard Bay. This mudstone exhibits a 'flow' fabric, and was thought by Thomas (1977) to be laharic in origin. A conformable andesite 2.5 m thick, which is present within the mudstone, is distinct in that it contains no phenocrysts. Just to the north of this andesite a vertical junction between mudstone and ignimbrite is exposed [6646 5567]. The ignimbrite here has the rubbly, brecciated texture characteristic of the base of a flow, and the cross-cutting relationship confirms that the boundary is unconformable.

Conglomerates on Long Echet (formerly called the Long Echet Conglomerate) and in the south-east corner of Giffard Bay are now both termed L'Homme Mort Conglomerate. They overlie andesite and the Jersey Shale Formation in unconformable relationship. Pebbles of andesitic and sedimentary rock types are predominant, with lesser amounts of granite, porphyry, rhyolite, vein-quartz, and metamorphic rock. The rhyolitic debris resembles material in the Long Echet Tuff (p.17), rather than in the main rhyolite sequence, and it is likely that this conglomerate is older than the main acidic volcanism. The L'Homme Mort Conglomerate is cut by the Frémont Fault, whereas the nearby outlier of Rozel Conglomerate is not, showing that the two deposits are of different ages.

The St John's Rhyolite Formation includes five ignimbrite cooling units, which are distinguishable by their field appearance. These ignimbrites have internal base-surge horizons, pumiceous horizons, and a vertical variation in

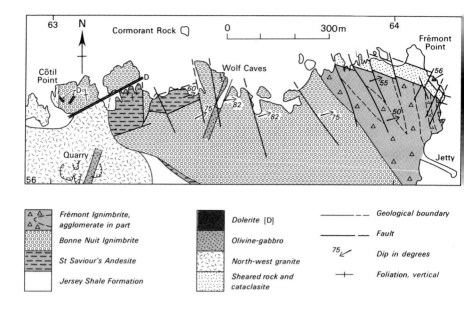

Legend:

Frémont Ignimbrite, agglomerate in part	Dolerite [D]
Bonne Nuit Ignimbrite	Olivine-gabbro
St Saviour's Andesite	North-west granite
Jersey Shale Formation	Sheared rock and cataclasite

— — — Geological boundary

——— Fault

75⤲ Dip in degrees

—⊢— Foliation, vertical

Figure 7
Geological sketch map of the coast between Côtil Point and Frémont Point. Based on Thomas, 1977, fig.6.2

phenocryst content, all indicating that they were compound cooling units containing numerous individual flows. In many cases erosional features are present between flows in a single ignimbrite. All the ignimbrites display eutaxitic textures and many individual fiamme show axiolitic devitrification. Corroded quartz phenocrysts and perthite laths are ubiquitous, and albite laths are abundant in the two oldest ignimbrites, the Jeffrey's Leap Ignimbrite and the Bonne Nuit Ignimbrite. All the ignimbrites contain fragments of country rock derived from either the magma chamber or the conduit and concentrated in base-surge horizons; shale and andesite are common, acidic volcanic material generally rare. In the Jeffrey's Leap Ignimbrite, assimilation of andesite has led to the growth of amphibole and biotite phenocrysts. An abundance of aphyric rhyolitic fragments in the Anne Port Ignimbrite, the last member of the St John's Rhyolite Formation on the east coast, demonstrates that such material had previously been erupted, but no outcrop is known.

North coast The St John's Rhyolite crops out along the north coast from Côtil Point to La Crête. The sequence between Côtil Point and Frémont Point (Figure 7) has been displaced by a sinistral tear fault at the south-west corner of Bonne Nuit Bay and is repeated between Bonne Nuit jetty and La Crête.

The Bonne Nuit Ignimbrite (Figure 8), at the bottom of the succession, is grey to black where fresh, and weathers to a pale pinkish cream colour. The base is only exposed in the

Figure 8 Variation in the lithology of the Bonne Nuit Ignimbrite, shown as a reconstructed vertical section at Bonne Nuit Bay. Redrawn from Thomas, 1977, fig.6.5

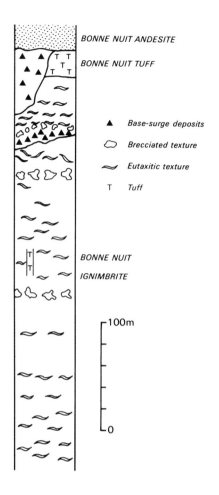

BONNE NUIT ANDESITE

BONNE NUIT TUFF

▲   Base-surge deposits

◯   Brecciated texture

〰   Eutaxitic texture

T   Tuff

BONNE NUIT IGNIMBRITE

100m

0

aureole of the north-west granite. Andesite blocks are common near the base and are generally aligned parallel to it. In higher levels the ignimbrite is characterised by many angular siltstone xenoliths, few of which exceed 20 mm in diameter; quartz, perthite, and albite phenocrysts are present in variable amounts. Fiamme make up 3 to 10 per cent of the rock; they are no longer glassy, some having been replaced by trails of granular quartz and feldspar, others displaying axiolitic devitrification. Several autobrecciated horizons are believed to represent the tops of individual flows. The thickness of the Bonne Nuit Ignimbrite is estimated to range from 550 m to 900 m between Côtil Point and Frémont Point, but in Bonne Nuit Bay the thickness varies from 400 m to 650 m and field evidence indicates that the upper surface was deeply eroded.

The Bonne Nuit Tuff, comprising light grey fine-grained tuff, overlies the Bonne Nuit Ignimbrite, and is succeeded by andesitic lavas and agglomerates. A lower agglomerate is up to 65 m thick and can be divided into three layers on the basis of its contained andesite fragments; the andesite in the lowest layer has pink albitised plagioclase phenocrysts, that in the middle layer is aphyric and microlitic, and the andesite in the uppermost layer has white phenocrysts. The matrix is felsitic. These are thought to be base-surge deposits. The Bonne Nuit Andesite, 60 m thick, overlies the pyroclastic rocks; plagioclase phenocrysts become fewer and smaller upwards, and the top is amygdaloidal. The andesite is in turn followed by 50 m of agglomerate, the Bonne Nuit Agglomerate, in which andesite fragments up to 1.5 m across are set in a fine-grained felsitic matrix rich in small angular quartz grains. Inland, and in the coastal section between Côtil Point and Fremont Point, the Bonne Nuit Ignimbrite is directly overlain by the Frémont Ignimbrite, and it is probable that the intervening andesitic rocks were confined to an erosional canyon a few hundred metres wide in the Bonne Nuit Ignimbrite.

The Frémont Ignimbrite is 160 m thick at Frémont Point. It rests on an irregular eroded surface and its top is not exposed. An uneven basal zone up to 15 m thick contains many andesite blocks which are up to 2 m across. The ignimbrite above can be divided into five units; four of these are individual flows and the fifth consists of several thin flows. Each flow is rich in xenoliths near its base but xenolith-poor and eutaxitic near its top. The xenoliths include shale, andesite, tuff, porphyritic rhyolite, aphyric rhyolite and agglomerate. The two lowest units have lenticular outcrops and they probably filled a pre-existing valley. At La Crête a maximum of 100 m of Frémont Ignimbrite is exposed, in a westward-dipping block separated from the eastward-dipping sequence to the south by a NW–SE fault; this block is deemed to be overturned on the evidence of the distribution of its xenolithic and eutaxitic horizons.

At Vicard Bay, north of the Frémont Fault, the Vicard Mudstone and Vicard Tuff are succeeded unconformably by ignimbrite 7 m thick, followed by 13 m of sedimentary breccia and by a further 65 m of ignimbrite. Both ignimbrites are grey-green and fine grained, and resemble the Frémont Ignimbrite. The sedimentary breccia consists of angular fragments of green shale and green tuff in a green fine-grained matrix of quartz and clay minerals; this breccia is probably an inter-flow scree-type deposit.

East coast Along the coast between Petit Portelet [716 504] and Anne Port [714 508] the foreshore reefs are formed of Jeffrey's Leap Ignimbrite; the limits of this member are obscured by beach deposits but the thickness exposed is about 200 m. The groundmass of the ignimbrite is generally felsitic, with grain-size less than 0.01 mm, but in places poorly defined areas show much coarser felsitic devitrification. Fiamme are generally difficult to discern, but where present they pick out a eutaxitic texture which dips at about 40° to just east of north. Xenoliths are abundant in areas poor in fiamme, and consist of sedimentary rocks probably derived from the Jersey Shale Formation, andesite, and salmon-pink porphyry containing albite, perthite and quartz

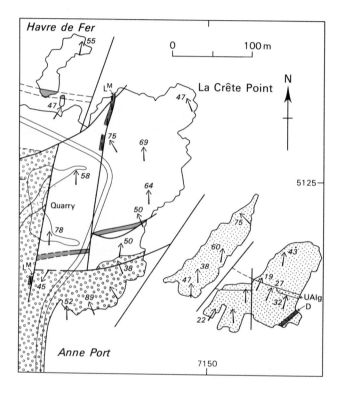

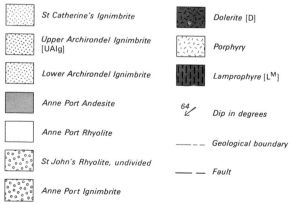

Figure 9 Sketch map of the geology north of Anne Port. Redrawn from Thomas, 1977, fig. 7.7

St Catherine's Ignimbrite

Upper Archirondel Ignimbrite [UAIg]

Lower Archirondel Ignimbrite

Anne Port Andesite

Anne Port Rhyolite

St John's Rhyolite, undivided

Anne Port Ignimbrite

Dolerite [D]

Porphyry

Lamprophyre [LᴹM]

64 Dip in degrees

Geological boundary

Fault

phenocrysts in a felsitic matrix; the largest shale and andesite xenoliths and most of the porphyry blocks are concentrated in the top 20 m or so, making a distinctive band. Phenocrysts of quartz, perthite, albite and altered mafic minerals occur throughout this ignimbrite; the mafic minerals may have been biotite and amphibole.

On the northern side of Anne Port (Figure 9) the Anne Port Ignimbrite is exposed. The flow is about 50 m thick. The lowest part visible is eutaxitic and passes up into almost parataxitic ignimbrite containing large xenolith-rich patches. Few of the fiamme are axiolitic. The top of the flow is rich in undeformed shards and pumice. The groundmass is fine grained and felsitic, and is stained and replaced by hematite, particularly towards the top of the flow. The combination of secondary hematite and microcrystalline quartz has given rise to jasper. Bipyramidal quartz and perthite phenocrysts in the ignimbrite vary from euhedral to angular or corroded. Xenoliths of andesite, aphyric rhyolite and ignimbrite are common and generally small. Within the central part of the flow several agglomeratic patches with sharply defined margins have many xenoliths set in a felsitic matrix free of fiamme and shards.

Inland exposures Ignimbrite in Queen's Valley [6935 4964 to 6959 4969] is indistinguishable from the Jeffrey's Leap Ignimbrite. It may be that the elongated outcrop of ignimbrite that extends from Queen's Valley to south of Le Bourg represents a fissure which fed the ignimbrite eruption.

An old quarry [6937 5110] south of Le Côtil Farm is in porphyritic ignimbrite containing a few small sedimentary xenoliths, and albite and quartz phenocrysts; perthite phenocrysts are abundant and eutaxitic texture is well developed. This rock resembles the Bonne Nuit Ignimbrite, which is also commonly exposed between St Manelier [674 510] and Augrès Mill [652 515], and along Grands Vaux and Vallée des Vaux.

A distinctive purple tuffaceous rock crops out on the north-east side of Grands Vaux between Trinity playing fields [660 535] and Rue Coutanche [654 546], and also along Trinity Road [6610 5375]. The location of this rock, the Trinity Ignimbrite, suggests that it is younger than the Frémont Ignimbrite, but it cannot be correlated with any other member of the St John's Rhyolite Formation.

Bouley Rhyolite Formation

Ignimbrites in the Bouley Rhyolite Formation around Bouley Bay are similar to the aphyric material from the east coast, but they are thicker, and several are demonstrably compound cooling units composed of two or three separate

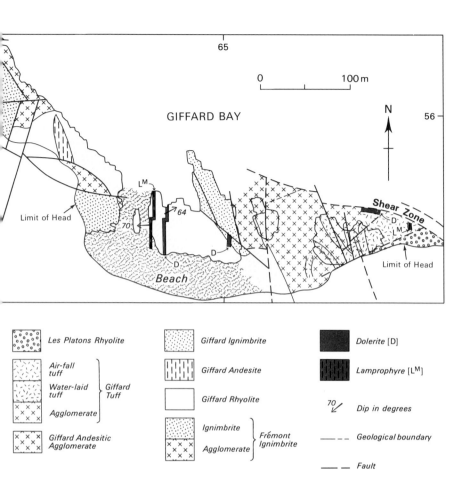

Les Platons Rhyolite

Air-fall tuff / Water-laid tuff / Agglomerate } Giffard Tuff

Giffard Andesitic Agglomerate

Giffard Ignimbrite

Giffard Andesite

Giffard Rhyolite

Ignimbrite / Agglomerate } Frémont Ignimbrite

Dolerite [D]

Lamprophyre [L^M]

70 Dip in degrees

Geological boundary

Fault

Figure 10 Sketch map showing the geology on the foreshore at Giffard Bay. Based on Thomas, 1977, fig.6.13

flows. Textures in all these ignimbrites are similar to those of the older ignimbrite flows, but xenoliths are acidic rather than andesitic or sedimentary, and spherulitic devitrification is commonly spectacular.

On the north coast, in Giffard Bay (Figure 10), a thin porphyritic ignimbrite and a thick sequence of pyroclastic rocks are sandwiched between rhyolite flows. Some of the pyroclastic rocks were deposited in pools of standing water and some were reworked by streams.

Giffard Bay The boundary between the St John's Rhyolite Formation and the Bouley Rhyolite Formation is exposed on the south-east corner of La Crête headland. At the base of the overturned succession (Figure 11) agglomerate is followed by the Giffard Rhyolite with an irregular eroded contact; the basal rhyolite is about 40 m thick, autobrecciated at the top, and followed by the thin (8 m) Giffard Andesite and then more rhyolite. This sequence is separated by a fault

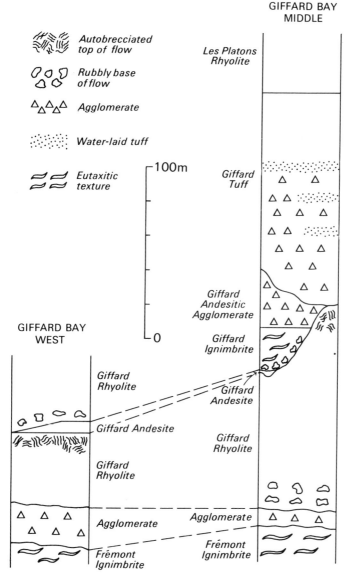

Figure 11
Generalised vertical sections through the rocks exposed in foreshore reefs at Giffard Bay, showing local absence of some formations and several erosion surfaces. Based on Thomas, 1977, fig.6.14

Legend:
- Autobrecciated top of flow
- Rubbly base of flow
- Agglomerate
- Water-laid tuff
- Eutaxitic texture

GIFFARD BAY MIDDLE

Les Platons Rhyolite

100m

Giffard Tuff

Giffard Andesitic Agglomerate

0

Giffard Ignimbrite

Giffard Andesite

GIFFARD BAY WEST

Giffard Rhyolite

Giffard Andesite

Giffard Rhyolite

Agglomerate

Frémont Ignimbrite

Giffard Rhyolite

Giffard Andesite

Giffard Rhyolite

Agglomerate

Frémont Ignimbrite

from the succession in the middle of Giffard Bay, where the dip is to the east; the lower rhyolite is about 100 m thick and the upper flow is missing through erosion or non-deposition. The andesite is preserved in small patches at the bottom of a contemporaneous gulley cut into the lower rhyolite. Later ignimbrite and andesitic agglomerate fill the gulley and the whole succession is overlain by a suite of acidic tuffs and agglomerates, and lastly by the Les Platons Rhyolite.

The Giffard Rhyolite is dark purple and fine grained, with contorted flow banding picked out in pale pink layers composed of outward-growing fans of fibrous feldspar. Trains of small pink spherulites parallel the flow banding, and small corroded megacrysts of quartz and perthite are present. The groundmass has undergone felsitic devitrification.

The Giffard Andesite is up to 8 m thick, dark grey and fine grained; it contains a few albitised plagioclase phenocrysts and albite microliths set in a felsitic matrix. The Giffard Ignimbrite succeeded the andesite in the gulley; it has a maximum thickness of 25 m and a flat upper surface. The matrix is fine grained and felsitic, and eutaxitic fiamme show axiolitic devitrification similar to that of the Frémont Ignimbrite. The base contains numerous xenoliths of rhyolite and andesite, and shale xenoliths and quartz and perthite phenocrysts are scattered throughout the flow. The Giffard Andesitic Agglomerate, also in the gulley, is 15 to 40 m thick. It consists of irregular blocks of porphyritic andesite, up to 0.4 m across, set in a fine-grained tuffaceous matrix which is felsitic, with small fragments of rhyolite and quartz. Casimir and Henson (1955) suggested that the deposit was produced when a flow of andesite broke up and became incorporated into an ash over which it was moving; Thomas (1977) preferred to call it 'Agglomeratic Andesite'.

Air-fall pyroclastic debris, partly water-laid, followed the Giffard Rhyolite and the Andesitic Agglomerate with a sharp base in western Giffard Bay, and constitutes the Giffard Tuff. Green air-fall tuff at the base is followed by purple mudstone, agglomerate, tuff, purple mudstone, and agglomerate. The purple mudstones were interpreted by Casimir and Henson (1955) as blocks of Precambrian sediment incorporated in the agglomerate, but Thomas (1977) regarded them as lacustrine sediments deposited on the irregular surface of the air-fall tuff, with local mixing and slumping. The agglomerates in the sequence contain mudstone, andesite and rhyolite debris, as well as albite, perthite, quartz, pumices, and shards, all set in a pale green fine-grained felsitic matrix. Towards the top of the succession the agglomerates grade into air-fall tuffs which have a similar matrix but fewer and smaller lithic and crystalline fragments and more pumiceous material.

Fluvially deposited volcaniclastic rocks of limited spread and thickness occur in the lower agglomerates. The most extensive of these is 35 m above the base of the pyroclastic sequence, 30 m across and 5 m thick; graded units are up to 0.3 m thick, and purple mudstones up to 15 mm thick occur at the top of most. Reverse grading, braiding, ripple marking, and contemporaneous microfaulting and slumping, all testify to deposition in a high-energy environment, probably a braided stream.

Water-laid tuffs characterised by alternating green and purple colouration occur at four levels in the air-fall pyroclastic rocks. Pumice, shards, and minor lithic fragments are set in a fine-grained felsitic matrix. There is some grading of pumices, but there appears to be no relationship between grading and colour. Accretionary lapilli that occur in some of the purple bands may be the 'elliptical . . . bombs' described by Casimir and Henson (1955). The shapes of these bodies of banded tuff suggest that they were deposited in pools of standing water.

The pink Les Platons Rhyolite overlies the tuffs in the south-east corner of Giffard Bay with a sharp junction. The basal 5 m has poor banding and contains pieces of tuff; flow banding is well developed above this. Spherulites up to 30 mm across are concentrated in bands parallel to the flow; they are composed of radiating fans of fibrous feldspar separated by films of hematite. The few pink perthite megacrysts present in the rock usually form the cores of spherulites. A brecciated horizon, 3 m thick, indicates that more than one flow occurred. The matrix is granular quartz with small open feldspar spherulites.

Bouley Bay In north-west Bouley Bay the reddish Vicard Point Ignimbrite is about 25 m thick. It contains several rubbly horizons and has an autobrecciated top; these combine with the flow-banded appearance of the rest of the unit to suggest a number of rhyolite flows. The overlying Lower Bouley Ignimbrite is about 80 m thick; its pale green colour is distinctive, but pink patches around Bouley Bay jetty are due to disseminated secondary hematite. Shards, pumice, crystal and lithic fragments, and spherulites are set in a fine-grained felsitic matrix. Locally – for example, at the roadside [6692 5468] 150 m south of Bouley Bay jetty – concentrations of spherulites form up to 80 per cent of the rock, their bright red colour contrasting strongly with the green matrix (Plate 4). The spherulites are built up of concentric shells of radiating fibrous feldspar formed during successive stages of growth and separated by granular quartz in some cases. Spherulites reaching 10 cm in diameter are present in the crags of Les Hurets; these spectacular rocks have been described by Mourant (1932; 1933). Spherulites and perlitic cracks are high-temperature features which distinguish the deposit as an ignimbrite. Above the Bouley Bay Ignimbrite, the Les Hurets Tuff is found only on the west side of Bouley Bay [6688 5463], where it is up to 35 m thick. It is distinguished by corroded bipyramidal quartz and pink perthite phenocrysts, and by a lack of xenoliths. Fiamme seen in thin section demonstrate its ignimbritic nature.

The reddish purple Middle Bouley Ignimbrite is exposed at the shoreline and on the heights above west Bouley Bay, at L'Islet, and at La Tête des Hougues. At the first and third of these localities numerous patches of large spherulites are set

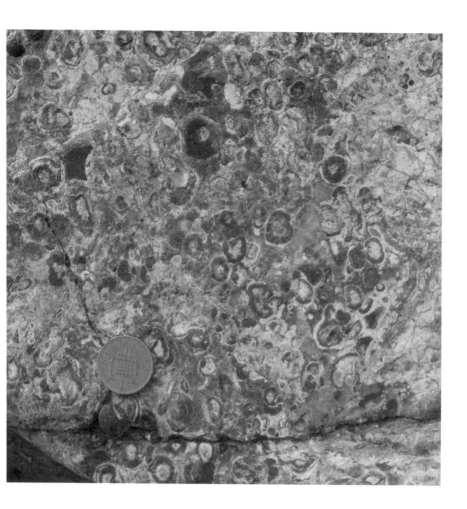

in brecciated material with a poorly defined eutaxitic texture, and the base is commonly brecciated. At L'Islet the ignimbrite displays superb parataxitic texture (Plate 5), giving it the appearance of a flow-banded 'rhyolite'; the underlying green Lower Bouley Ignimbrite has been affected by this intense welding. This spatial distribution suggests a section across a flow, the margins having cooled more rapidly than the interior, producing primary spherulitic growth, and the middle being more strongly welded.

Salmon-pink Upper Bouley Ignimbrite lies above the darker Middle Bouley Ignimbrite at La Tête des Hougues and at L'Islet, and can also be recognised in isolated exposures south-west of Bouley Bay jetty. It is up to about 200 m thick and consists of two flows. The lower of these has a brecciated base and top, and is highly welded and compacted near its base. The main part of the flow is eutaxitic, with fiamme ranging from 0.05 to 30 mm in length and

roughly one tenth as thick, and feldspar xenocrysts replaced by sericite and microcrystalline quartz. The eutaxitic ignimbrite gives way upwards to pumiceous ignimbrite. The upper flow has a sharp, undulatory, brecciated base which shows little compaction. Parataxitic textures are present 25 m above the base. Fiamme reach 0.4 m in length, giving a flow-banded appearance. Most of the groundmass has undergone spherulitic devitrification. The top of the flow is not seen.

Plate 5 Parataxitic texture in the Middle Bouley Ignimbrite at L'Islet, Bouley Bay. (A13683)

East coast The Anne Port Rhyolite is a sequence of five rhyolite flows which overlies the Anne Port Ignimbrite on an irregular, slightly eroded surface (Figure 12). In the ideal flow a rubbly base passes upwards into massive flow-banded lava with columnar jointing (Plate 6), and the top is autobrecciated, but inter-flow erosion has removed the upper part of some of the flows. Of the five flows, the top four are present in Anne Port Quarry [713 512] and roadside exposures to the south of it. The Anne Port Andesite provides a thin marker horizon between the third and fourth flows. The columnar uppermost rhyolite is repeated by faulting in Havre de Fer, its top being hidden by the Archirondel breakwater.

The rubbly base of each flow contains numerous angular rock and mineral fragments up to 2 m across, set in a fine-

Figure 12 Vertical section through the Anne Port Rhyolite exposed on the north side of Anne Port beach. Flow 5 is not drawn to scale. Based on Thomas, 1977, fig.7.14

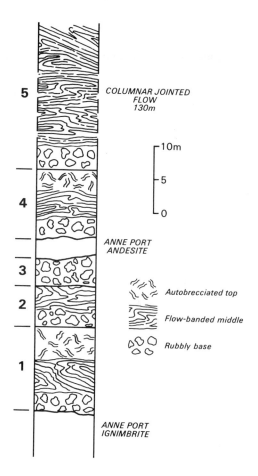

5   COLUMNAR JOINTED FLOW 130m

10m

5

0

4

ANNE PORT ANDESITE

3

2   Autobrecciated top

Flow-banded middle

Rubbly base

1

ANNE PORT IGNIMBRITE

grained felsitic matrix patchily rich in finely divided secondary hematite. Flow banding is picked out by an alternation of dark purple spherulitic bands and pink quartz-rich bands. Many of the purple bands display a continuous median layer of magnetite.

The Anne Port Andesite is similar to the lavas in the St Saviour's Andesite Formation; it also resembles the Giffard Andesite, with which it is tentatively correlated. Its lower and upper surfaces are irregular, and its thickness ranges from 1 to 3 m. The andesite is dark grey and fine grained, with albite phenocrysts and chlorite pseudomorphs after an unidentified mafic constituent, set in a matrix of microlites, magnetite and feldspar.

The succession north of the Archirondel breakwater is folded about E–W axes (Figure 13). Lower Archirondel Ignimbrite crops out just north of the breakwater, which conceals its base. The salmon-pink aphyric ignimbrite is about 55 m thick. Eutaxitic texture is present everywhere, but par-

ticularly in exposures farther north [7094 5204]. Macroscopic pink spherulites, which weather white, occur in patches, and quartz and perthite megacrysts are randomly scattered in the devitrified matrix.

Near the Dolmen* [7092 5197] the Lower Archirondel Ignimbrite is succeeded by pink Lower Archirondel Tuff, the exposed thickness of which is up to 1.95 m. The base is concealed by beach deposits and the top is irregular. Pale green to white pumice and rhyolite, andesite, quartz and albite fragments are set in a fine-grained felsitic matrix, much of which has been replaced by hematite from which it derives its colour. Well-developed sorting distinguishes these deposits

Plate 6 Columnar jointing in the Anne Port Rhyolite at Anne Port. Notebook for scale. (Photograph by Dr D. G. Helm)

* Dr J. T. Renouf has pointed out that the 'Dolmen' marked on the topographical base-map is in fact an 18th century gun platform.

Figure 13 Sketch map showing the geology north of Archirondel Tower. Based on Thomas, 1977, fig.7.16

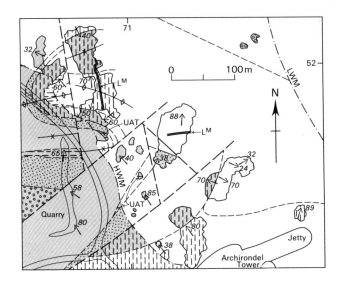

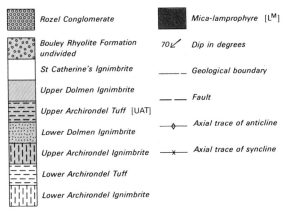

	Rozel Conglomerate		Mica-lamprophyre [L^M]
	Bouley Rhyolite Formation undivided	70↙	Dip in degrees
	St Catherine's Ignimbrite	_____	Geological boundary
	Upper Dolmen Ignimbrite		
	Upper Archirondel Tuff [UAT]	__ __	Fault
	Lower Dolmen Ignimbrite		
	Upper Archirondel Ignimbrite	◇	Axial trace of anticline
	Lower Archirondel Tuff	×—	Axial trace of syncline
	Lower Archirondel Ignimbrite		

as air-fall tuffs rather than flow-pyroclastics, and the thin fine-grained graded upper parts of some beds suggest post-eruptional dust settling above an air-fall deposit, rather than water-sorted debris.

Elsewhere the Lower Archirondel Ignimbrite is directly followed by the purple Upper Archirondel Ignimbrite, which is up to 28 m thick. The basal 5 m or so are rubbly, and rich in pumice and rhyolite blocks, some of which are more than 1 m in diameter. The rubbly base passes up into fragment-rich eutaxitic ignimbrite where the flow is thicker, but it is followed by autobrecciated material where the flow is thinner. Spherulites up to 5 mm in diameter, showing up to six growth stages, are set in granular quartz showing perlitic cracking. The matrix varies from felsitic to spherulitic, eutaxitic texture is poor, and fiamme, which do not exceed

10 by 0.7 mm, provide the only indication that this is an ignimbrite rather than a lava flow.

The salmon-pink aphyric Lower Dolmen Ignimbrite overlies the Upper Archirondel Ignimbrite; it thickens southwards from 4 m on the shore [7095 5194] to 30 m in a quarry [7090 5183] 250 m north-west of Archirondel. The base is rubbly and the top autobrecciated. Aligned pumice fragments up to 80 mm long form up to 40 per cent of the rock and show little evidence of compaction. These characteristics suggest that the deposit was originally a nonwelded ignimbrite.

The Upper Archirondel Tuff follows the Lower Dolmen Ignimbrite on the shore [7096 5190] and in the quarry [7090 5189], but it is impersistent, ranging up to 14 m in thickness. It is a purple fine-grained rock, rich in pink perthite and quartz phenocrysts, and has the appearance of a porphyry, but field evidence suggests that it was extrusive. Perthite and albite laths form clusters 10 mm across and show corrosion and internal brecciation, but little sericitisation; together with embayed cracked bipyramids of quartz, and what may be fiamme of unusual shape, they are set in a felsitic groundmass rich in finely divided magnetite and secondary hematite. The poorly axiolitic eutaxitic texture and the corroded state of the phenocrysts suggest that this rock may be an ignimbrite.

Where the Upper Archirondel Tuff is absent the purplish pink aphyric Upper Dolmen Ignimbrite follows the Lower Dolmen Ignimbrite, fragments of the latter being caught up in the overlying flow. The Upper Dolmen Ignimbrite has a rubbly base about 4 m thick, in which rhyolite fragments are set in a fine-grained felsitic matrix. The median part of the flow shows poor columnar jointing and fine fiamme give the rock a flow-banded appearance. Pink spherulites up to 4 mm in diameter occur in aligned patches a few metres long, which contain no fiamme.

The St Catherine's Ignimbrite is probably the youngest of the Jersey volcanic rocks. At Anne Port [7160 5112] it dips to the north over an irregular surface, below which a near-vertical boundary between Lower and Upper Archirondel Ignimbrites can be seen, and north of Archirondel Tower it rests on Upper Archirondel Ignimbrite in one exposure [7114 5183] and on Lower Dolmen Ignimbrite in another [7107 5189]. Thus considerable erosion occurred prior to the deposition of the St Catherine's Ignimbrite.

At Anne Port the lowest 5 m of the St Catherine's Ignimbrite are rich in ellipsoidal black fiamme, which reach 0.5 m diameter and 25 mm thickness. This horizon has a distinctive green fine-grained groundmass and pink perthite phenocrysts, and is not seen elsewhere. The rest of the ignimbrite (80 m) is characterised by eutaxitic texture and by a dull grey flinty groundmass. Fiamme rarely exceed 0.2 m in length. Xenoliths of pink ignimbrite (possibly Lower Archirondel Ignimbrite), up to 26 m by 3 m, are aligned near the top of the flow. Corroded shattered bipyramidal quartz and perthite phenocrysts reach 3.5 mm in diameter.

Rozel Conglomerate Formation 4

The Rozel Conglomerate Formation comprises coarse conglomerates with subordinate sandstones and mudstones. The main outcrop of the formation occupies the north-eastern corner of Jersey and forms the lowest of the cliffs along the northern coast of the island; the inland junction with the Bouley Rhyolite Formation is mainly concealed by superficial deposits. Two outliers are situated to the west, one in Les Hurets valley [665 548] to the west of Bouley Bay, and the other farther north, on the Pierre de la Fételle heights [663 556] above Vicard Point.

The distinctive pebbly lithology of this formation (Plate 7) caught the observant eye of Philip Dumaresq as early as the 1680s (Dumaresq, 1685), though it was not until the second half of the 19th century that the formation was fully described and its main boundaries were defined with any precision (Ansted and Latham, 1862; Noury, 1886). Noury (1886, pp.135–136) discussed whether the deposits were of marine or continental origin; he favoured the former, and it was not until Cornes (1933) published his account that the continental nature of the formation was established to the extent that this remains the accepted origin (Squire, 1970; Renouf, 1974; Duff, 1979).

The Rozel Conglomerate is indisputably the youngest of the major hard rock formations in Jersey, but its age in terms of the geological column is uncertain. Recent work (e.g. Adams, 1976; Duff, 1979, 1980) has narrowed the field of choice from a Precambrian to Permo-Carboniferous range to one spanning Cambrian to Siluro-Devonian time, with the emphasis focusing on late Cambrian to early Ordovician (e.g. Squire, 1970; Renouf, 1974; Duff, 1979, 1980). Duff (1979, p.359) suggested that the oldest stable natural remanent magnetisation (NRM) component of the mudstones in the sequence (p.42) was acquired during authigenesis of hematite in Silurian or early Devonian times, but that this was not incompatible with a Cambro-Ordovician age for the rocks themselves. A possible palaeontological indication of age has been given by B. H. Bland (1984), with his assignment of structures previously identified as raindrop and related prints in the mudstone below La Tête des Hougues (see p.41) to the large impression fossil *Arumberia* Glaessner & Walter. This fossil has so far been found in rocks of latest Precambrian to Lower Cambrian age, but its full range has

Plate 7 Rozel Conglomerate below La Tête des Hougues, Trinity. (A13692)

not been established. Adams (1976) gave a date of 427 ± 13 Ma (435 ± 13 Ma when recalculated) for a dyke of hornblende-lamprophyre [7110 5380] that has intruded the conglomerates just south of La Coupe Point; this Upper Ordovician/Lower Silurian date thus sets a minimum for the age of the conglomerates.

The base of the Rozel Conglomerate is exposed on the cliff slopes above Vicard Point and in the small bay below La Tête des Hougues. There was formerly an exposure of the base of the Les Hurets valley outlier on Bouley Bay hill [6638 5456] but this is now obscured. At all three localities the conglomerates rest on rhyolite; their bedding dips into or against the volcanic rock, indicating that the pre-conglomerate surface was eroded, the depressions being at least of the order of size of the Les Hurets valley itself.

Above Vicard Point the conglomerate [6645 5563] dips south-westward into the hillside at between 30° and 40°; the surface of the Bouley Rhyolite is visibly uneven, with northward-trending, smooth-sided channels of the order of several metres in length and up to 1 m deep (Thomas, 1977). Contours drawn on the base of the conglomerate indicate that the outlier fills a strongly defined hollow cut in the rhyolites. Where visible the basal beds are conglomerates, and lack the finer-grained beds such as are to be found in Les Hurets valley and in the main outcrop.

In the little bay [679 545] below La Tête des Hougues the conglomerates rest on both the Middle and the Lower Bouley Ignimbrite. The contact can be traced from the waterfall at the back of the bay, along its western side and then eastward for a short distance at low tide. There is a visible 2 to 3 m relief to the underlying surface of the volcanic rocks which has been revealed by recent erosion, and the presence of an isolated patch of conglomerate beside the path that descends to the modern beach emphasises the irregularity of the surface. The thickness of strata in the main outcrop is estimated at 500 to 800 m. Overall, the conglomerates are coarse-grained, poorly sorted rocks, which were deposited in units mostly upwards of 1 m thick. Boulders up to 2 m in diameter are found in some places, for example at and near the Tour de Rozel [6918 5520; 6924 5508]. Bands of coarse grit occur here and there throughout the exposed succession [6792 5448; 6924 5508; 7114 5288], though none has been noted in either of the outliers. More localised are small thicknesses of much finer-grained beds containing maroon mudstone bands a few centimetres thick [6792 5448; 6934 5440; 704 520; 7114 5288]. A lack of recognisable marker horizons has so far prevented the subdivision of the formation.

Below La Tête des Hougues the mudstones and associated finer-grained beds in the conglomerate succession show graded bedding (Plate 8) and mud cracks, and these indicate that the sediments were deposited in shallow pools, open to small influxes of muddy water and subject to drying out. Mourant (1933) noted 'impressions of raindrops, rill or ripple-marks' in these mudstones, but B. H. Bland (1984) has suggested that these structures represent the trace fossil *Arumberia* (see p.40); Bland considered that though 'a braided stream or an intertidal sand and mudflat' was the most likely environment of deposition for *Arumberia*-bearing strata, 'shallow lakes in a flood-plain' were not ruled out. The mudstones below La Tête des Hougues were originally about horizontal but they now dip at about 36° to the northeast, as do the main conglomerate beds, thus showing that the dip has been structurally imposed. Unfortunately the other mudstones in the succession are too localised to do more than hint at the overall depositional development of the formation. The two principal mudstone outcrops, below La

Plate 8
Mudstones near
the base of the
Rozel
Conglomerate
below La Tête des
Hougues. Each
unit has a
conglomeratic
base that grades
up to a mudstone
top. (A13693)

Tête des Hougues and opposite La Solitude Farm [7042 5202], are characterised by oxidation (maroon) and reduction (green) hues, though the full significance of this as regards the climate at the time of deposition has not been resolved.

The localised occurrence of the mudstones and the evidence they present of deposition in shallow surface pools agree with the overall picture of a continental deposit built up from floods of unsorted material brought down from a hilly or mountainous terrain by rivers. Imbrication of the flatter pebbles is common, and Squire (1970) made a preliminary study of the source direction, finding a predominantly northerly origin, though with one ESE result that has not been confirmed. Thomas (1977) noted imbricate structure at La Tête des Hougues [679 545] and confirmed the derivation from the north. However, Helm (1984) and Richardson (1984) have drawn attention to tectonic flattening and pebble rotation in the Rozel Conglomerate (see p.76), which cast some doubt on the validity of inferences drawn from pebble imbrication as an indicator of palaeocurrent directions in this instance. Coarse cross-bedding is exposed at a number of localities and is well displayed between the Nez du Guet [6971 5479] and the foreshore outcrops immediately east of Rozel jetty. The polygenetic nature of the pebbles in the conglomerates has long been recognised. The chief groups of rocks represented are Brioverian metasediments, volcanic formations, and various coarse-grained acid plutonic rocks. The acid plutonic rocks do not closely resemble those cropping out in Jersey, nor have foliated granitic rocks like those of Les Ecréhous been found, but the Brioverian sediments and volcanic rocks match perfectly. The pebbles of acid plutonic rocks are generally more rounded than the other constituents, suggesting that they may have been derived from farther afield, though probably from no more than 10 to 20 km away. It appears that the local Cadomian granites and the associated more basic rocks had not been unroofed at the time when the Rozel Conglomerate was deposited. Thomas (1977 and personal communication) recorded the occurrence of cobbles of conglomerate at La Tête des Hougues [679 545], which indicate that the deposit has been reworked; he also noted that volcanic material was common near the base, but rare or absent in the main body of the formation.

The matrix of the conglomerates is made of clay-size particles, but rarely exceeds 5 per cent of the total rock. There is little or no silt- or sand-sized fraction. The cement was originally hematitic, but this has been oxidised to limonite near the surface. The very localised occurrence of small reduction layers associated with the maroon mudstones has already been mentioned above.

Despite prolonged searches of the outcrops by many geologists, no trace of macrofossils other than the possible *Arumberia* (already mentioned) has been found, either as an original or as a derived constituent. The maroon mudstones and associated fine-grained beds have been exhaustively treated for the recovery of microfossils or organic traces, but with entirely negative results.

Plutonic igneous rocks 5

There are three principal plutonic rock complexes in Jersey, situated in the north-west, south-west and south-east of the island. In addition there is a smaller outcrop at Belle Hougue Point on the north coast, and a small, isolated, poorly exposed mass of diorite inland at Les Augrès in Vallée des Vaux, east of Becquet Vincent [647 521]. All the main masses comprise several granites which differ in texture, but the south-west complex is different from the others in lacking gabbroic and dioritic rocks. The plutonic complexes are post-tectonic, having been emplaced after the folding and low-grade regional metamorphism (low chlorite grade) of the Jersey Shale Formation and of the overlying volcanic rocks during the late Precambrian Cadomian earth movements.

Gabbro and diorite are everywhere older than the accompanying granites; nowhere in Jersey or the other Channel Islands are plutonic masses of diorite or gabbro intruded into granite, although some basic dykes, notably that at Wolf Caves [6347 5616], are gabbroic. The granites are of calc-alkaline type and are I-type granites (Chappell and White, 1974) typical of andinotype tectonic regimes. The granitic plutons disrupted the layered basic rocks, so that the latter are now preserved only as tilted remnants, some of considerable size.

Adams (1967, 1976) determined the isotopic ages of the Jersey granites, and with others (Bishop and others, 1975) discussed their regional setting. Duff's (1981) study of the palaeomagnetic patterns in the diorites showed complex magnetisation unrelated in a simple way to the original crystallisation period but probably resulting from recrystallisation and block rotation. Briden and others (1982; see also Chapter 9) have used geophysical data to show that the granitic complexes are continuous beneath the sedimentary and volcanic rocks of much, and probably all, of the island.

Gabbro and diorite

Gabbro and diorite are exposed near the eastern end of the north-west granite and extensively in the south-east complex. In addition, dioritic rocks make up a large part of the Belle Hougue mass, as well as the small outcrop at Les Augrès. The oldest plutonic rocks are layered gabbros (Figure 14), but most of the gabbros have been made over into diorites by interaction with the subsequent granites and by the metasomatic action of the fluids emanating from them. In the main diorite exposures in south-east Jersey, the layered diorites extend from the intertidal reefs at La Grève d'Azette to Havre des Fontaines, a distance which, taking account of the present dip of the layers (below) and possible repetition by later faulting, indicates an original layered sequence about 1 km thick. Further layered diorites occur near Seymour Tower [725 454], at the south-eastern tip of the island.

The contacts between gabbro and country rock have not been observed, apart from that of the Les Augrès mass. Exposures here are poor, but in the inter-war years a contact with andesitic rocks was visible and appeared not to be faulted (Dr A. E. Mourant, personal communication).

The original gabbroic mineralogy is preserved at Sorel Point [612 570] and, less extensively, at Le Nez Point [6783 4625] in the south-east complex; the layered structure is also preserved, though it is no longer in its original subhorizontal attitude. The layering at Sorel Point and Ronez is inclined at about 40° to the south, whereas at Le Nez Point the inclination is 70° or more towards the north-east, and near Seymour Tower the general dip of the layering is about 30°, also towards the north-east.

The gabbroic rocks at Sorel Point essentially comprise serpentinised olivine, clinopyroxene, and calcic plagioclase (labradorite), with accessory opaque minerals. Variation in the proportions of feldspar and pyroxene has produced layers that are, on average, about 1 m thick. Most of the rocks show progressive alteration of clinopyroxene, first to kaersutite and then to green hornblende, with an accompanying change of feldspar composition from labradorite to oligoclase/andesine. A little of the kaersutite may also be primary. The labradorite persists as cores of grains mantled with more sodic rims and, as the proportion of sodic plagiocase and hornblende increases, the texture changes from ophitic/poikilitic to one in which amphibole has increasingly idiomorphic outlines. Where recrystallisation reached this stage, the rocks became mobile and flowed, and megacrysts of potassic feldspar, in every way identical to those in the nearby granite, began to appear in the diorites. These neomagmatic rocks intruded their more rigid neighbours, resulting in a complex sequence of alteration, flow and intrusion.

From south-east Jersey, Bishop and Key (1983) have described another kind of layering – differentiated sheet layering – which is more widely distributed than gabbroic layering and has the same attitude and orientation. The differentiated sheet layering (Figure 14) comprises a succession of sheets, usually from 1 to 2 m thick, each formed of a lower

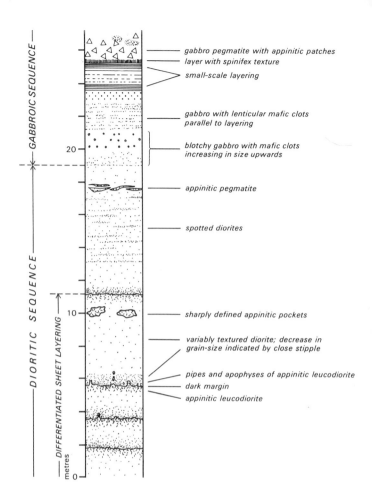

Figure 14 Vertical section through part of the layered gabbro-diorite sequence exposed at Le Nez Point. After Bishop and Key, 1983, fig.2

GABBROIC SEQUENCE

DIORITIC SEQUENCE

DIFFERENTIATED SHEET LAYERING

metres

gabbro pegmatite with appinitic patches
layer with spinifex texture
small-scale layering

gabbro with lenticular mafic clots
parallel to layering

blotchy gabbro with mafic clots
increasing in size upwards

appinitic pegmatite

spotted diorites

sharply defined appinitic pockets

variably textured diorite; decrease in
grain-size indicated by close stipple

pipes and apophyses of appinitic leucodiorite
dark margin
appinitic leucodiorite

dark member that grades upwards into more leucocratic diorite or quartz-diorite which is in sharp contact with the basal dark diorite of the overlying sheet (Plates 9 and 10). The uppermost, leucocratic quartz-diorites commonly contain slender amphibole crystals, many of which are aligned parallel to the contact. Numerous pipes, veins and apophyses of leucodiorite that penetrate upwards into the overlying dark diorite are usually no more than 30 mm across, and some are a metre or more long (Plate 11); they are all roughly perpendicular to the plane of the layering, and indicate that the meladiorite was sufficiently plastic to permit intrusion and that the rocks were in a near-horizontal attitude when these structures formed. Bishop and Key (1983) suggested that the differentiated sheet layers developed from gabbroic layers during dioritisation, and that progressive recrystallisation, mobilisation and flow led first to the

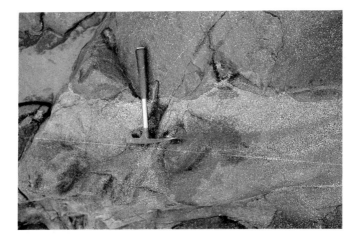

Plate 9
Differentiated
sheet layering in
diorite at Le Nez
Point, St
Clement.
(Photograph by
Dr A. C. Bishop)

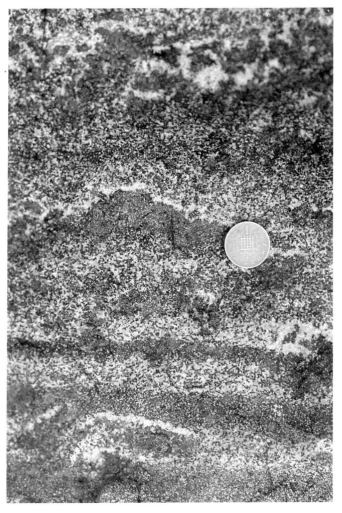

Plate 10 Thin
alternating layers
of pale and dark
diorite at Le Nez
Point, St
Clement.
(Photograph by
Dr A. C. Bishop)

Plate 11 A pipe of pale quartz-diorite in darker diorite, exposed in a reef near Seymour Tower. (Photograph by Dr A. C. Bishop)

modification of the sheets and later to the production of homogeneous diorites.

The Jersey diorites are of variable mineralogy, grain-size and texture. They show all gradations in composition from slightly altered gabbro, through hornblende-gabbro, to diorite and granodiorite. There are diorites which border on the lower limit of coarse-grained rocks but most are coarser than this, and pegmatitic diorites also occur. Most of the diorites are equigranular, with amphibole tending to show idiomorphic outlines against other minerals, particularly in the more siliceous and quartzose varieties. Some of the diorites, usually those of more basic composition, contain large equant poikiloblastic crystals of hornblende; such crystals are particularly abundant near Le Nez and at Ronez and Sorel points. Other diorites contain markedly elongate, prismatic amphiboles, many with cores of feldspar and quartz (Plate 12). Wells and Bishop (1955) drew attention to coarse-grained appinitic rocks of this kind at Le Nez Point, but the texture is typical of many leucodiorites and quartz-diorites, and it occurs in other mafic rocks as well. It is most probably the result of rapid crystallisation, possibly induced by loss of volatiles as a result of venting to the surface (Key, 1977).

Everywhere, veins of more felsic material abound in the diorites, and inclusions of diorite occur within more granitic rocks. These veins and inclusions are locally sharply bounded and show little or no evidence of reaction. Elsewhere they merge into the surrounding rocks, indicating that there was considerable interaction between the intrusive and host rocks. This veining and brecciation are most probably not the result of a single event, but of repeated episodes of intrusion.

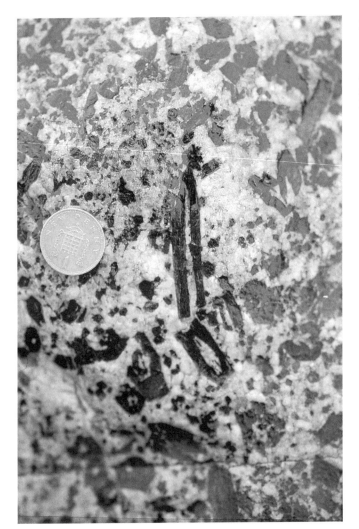

Plate 12 Skeletal amphibole crystals in appinite at Le Nez Point, St Clement. (Photograph by Dr A. C. Bishop)

Fine-grained diorites, particularly those which occur close to granite contacts, commonly contain ocelli, up to 5 mm or so across, of quartz, or quartz and potassic feldspar, rimmed by amphibole. Rocks containing these structures are known to the quarrymen at Ronez as 'bird's eye' and similar rocks abound in dioritic complexes elsewhere. Angus (1962), for example, described such rocks from Tyrone in Northern Ireland, and ascribed their origin to the crystallisation of quartz as a consequence of interaction of granitic magma and more basic rocks.

Most of the diorites appear not to be the products of primary crystallisation of dioritic magma, but to have been derived from gabbro. The alteration of gabbro resulted from changes in the composition of both felsic and ferromagnesian minerals in response to falling temperature. The change from calcic to sodic plagioclase has already been mentioned. Initially, cores of labradorite, generally clouded by minute rods of iron oxide, became mantled with oligoclase/andesine, which encroached on and replaced the cores, many of which were replaced by 'sericite' or prehnite. It is unusual to find a diorite with fresh feldspars; most have been sericitised to some extent. Parallel changes have affected the mafic minerals; the primary pyroxene of the gabbros has been replaced in a complex and intricate way by the brown, titanium-bearing amphibole kaersutite and by green hornblende.

Kaersutite itself was replaced by green amphibole – usually a magnesiohornblende – in places marginally and along cleavages, but more usually in an irregular fashion. In southeast Jersey it was commonly green amphibole that first replaced pyroxene, kaersutite being only rarely found. Here, however, as crystallisation of secondary amphibole proceeded and it outgrew pyroxene grain boundaries to form robust porphyroblasts, it was accompanied by a change in colour from green, though brownish green, to brown. This brown amphibole is a brown hornblende, however, and not kaersutite. More advanced alteration resulted in the patchy replacement of both brown and green hornblende by pale green actinolite and colourless tremolite, and by the crystallisation of chlorite, some of which also replaced hornblende and biotite directly.

At this stage, quartz became an important constituent and, in some rocks, was accompanied by potassic feldspar which usually occurs either as euhedral to subhedral pink megacrysts of orthoclase perthite, commonly mantled by a rim of white oligoclase in medium-grained diorites, or as an interstitial mineral, along with quartz, in coarser-grained diorites.

The accessory minerals of the gabbros and diorites include opaque iron and iron-titanium oxides; the latter probably contributed titanium to the kaersutite. Apatite occurs in two habits: it forms rather robust crystals in the gabbros, but is more prominent in the diorites, where acicular crystals are abundant. These elongated apatites span quartz/feldspar and feldspar/hornblende boundaries but are never found in the relict labradorite cores of feldspars; they appear to have formed during the dioritic recrystallisation.

The changes in composition of both the mafic and the felsic minerals resulted in an increase in the content of SiO_2 and CaO in the later rocks. As labradorite was progressively replaced by more sodic plagioclase, and as pyroxene was altered to amphibole, CaO was released. The appearance of sphene ($CaTiSiO_5$) as a common accessory mineral probably resulted in part from such changes. At low temperatures, when Ca-bearing amphibole was replaced by cummingtonite/anthophyllite and chlorite, further CaO must have been released, and such diorites contain many low-temperature Ca-Al silicates. The most common of these is epidote, much of which replaces feldspar or accompanies quartz as a late-stage cavity infilling. In many parts of Jersey, however, dioritic rocks have undergone more general epidotisation and have been partially replaced by epidote deposited from migrating fluids. Other low-temperature minerals which occur in diorites are prehnite, calcite and zeolites.

It is possible that the north-west and south-east granites were closely preceded by separate intrusions of gabbro, but the circumstantial evidence from the layering, the isotopic age determinations and the observation that nowhere in the Channel Islands has gabbro intruded granite argue that the major phase of gabbro emplacement was earlier.

Duff (1981) studied the natural remanent magnetisation (NRM) of the Jersey gabbros and found that the scattered but stable NRM could be accounted for by a model involving the disruption of a continuous, near-horizontal layered gabbro with a uniform NRM direction carried by titanomagnetite. The layered gabbro was intruded by granite, and during the ensuing dioritisation a variable, multicomponent NRM was imposed on the basic rocks, either by remagnetisation of the original gabbroic titanomagnetites or by the addition of a separate NRM component carried by magnetite which crystallised during dioritisation. Stoping, subsidence and rotation of masses of diorite within the granite caused dispersion of the pre-stoping NRM, and was the principal cause of the observed palaeomagnetic scatter. The Jersey gabbros, therefore, may once have formed part of a layered intrusion emplaced into Brioverian sedimentary and volcanic rocks and subsequently intruded by granitic plutons; it was probably related to and may have been joined to similar rocks in Guernsey and Alderney.

North-west granite

The north-west granite – comprising the granite of St Mary's type and the aplogranite of Mont Mado type – is the largest of the three principal masses in Jersey. Isotopic dating has shown it to be the youngest of them, with a Rb:Sr whole-rock isochron age of 480 Ma (recalculated from 490 Ma of Adams, 1976; Bishop and others, 1975). Intrusive contacts are exposed in several places: with sediments of the Jersey Shale Formation at L'Etacq [5580 5435] and Le Pulec [5475 5495]; with diorites in the area around Sorel and Ronez points; with rhyolite and andesite at the easternmost end of the outcrop in the vicinity of Côtil Point [631 562].

Hornfelsed Brioverian sediments are exposed close to the granite along its concealed southern and eastern margins; only in St Peter's Valley and at Handois does the margin appear to be faulted. In general the thermal effects of the granite are rather weak. The contact of granite and sediment is sharp, with relatively little brecciation, and there are few xenoliths of sediment in the granite. Veins of granite and aplite penetrate for some distance into the sediments, but there has been no extensive permeation by granitic fluids. Near L'Etacq [5580 5435] the granite makes a sharp contact with the sediments (Plate 1), the zone of thermal alteration being terminated at the La Bouque Fault. The more pelitic members of the Jersey Shale Formation are altered at the contact to cordierite-biotite-hornfelses, and thermal spotting is visible in some places.

The contact with volcanic rocks in the east is similar. At Côtil Point (Figure 7) sharply bounded veins of aplogranite have intruded ignimbrite and basic dykes within it. The granite and ignimbrite are so similar in composition that there is little evidence of thermal metamorphism, but the effects of K-metasomatism are shown by the presence of muscovite flakes coating joint surfaces in the ignimbrite. A little farther east [6318 5620] the granite has veined and brecciated the St Saviour's Andesite Formation, and garnets, probably of hydrothermal origin, have been recorded in the andesites in this area (Oliver, 1958).

The contacts between granite and sedimentary and volcanic rocks contrast markedly with those between granite and diorite. The former are sharp, simple and show little evidence of reaction; the latter are complex, and extensive reaction between granite and diorite is the rule. The greater reactivity of granite with diorite compared with other rocks cannot be ascribed simply to residual heat from the original crystallisation of the gabbro, for it appears that the gabbro was emplaced well before the oldest of the granites, the Dicq granite of south-east Jersey dated at about 570 Ma (see pp. 56–57). It is necessary, therefore, to argue either that the gabbros were especially prone to react with granite – a feature common in these rocks elsewhere – or that the gabbros associated with the north-west granite were emplaced separately from and later than those of south-east Jersey; the latter explanation seems unlikely.

Most of the north-west granite is coarse-grained – the St Mary's granite – consisting of rather tabular crystals of orthoclase or orthoclase perthite, with subordinate and smaller plagioclase, abundant quartz, and biotite and hornblende; zircon and apatite are common accessory minerals. Texture varies from place to place, but in general the St Mary's granite is of uniform appearance. From La Saline [6300

5615] to Les Mouriers Valley [605 562] the granite is a fresh pink, but along most of the rest of the north coast it is a rather greyish pink, owing to the presence of small, chloritised biotite crystals. It was formerly worked for road metal and building stone in several small quarries, notably at L'Etacq [563 542] and La Perruque [6280 5605]. Inclusions become noticeable in the granite east of Plémont Point [563 569], and are particularly abundant in the dioritic area around Ronez and Sorel points.

At the eastern end of the granite there is a marginal band, up to about 300 m wide, of aplogranite—the Mont Mado aplogranite. This is a handsome pink rock, poor in coloured minerals, and comprising orthoclase perthite and quartz. It was, in the past, widely used as a building stone, but the quarry at Mont Mado [637 556] has now been filled and the once-thriving industry has ceased, apart from a small quarry [6305 5602] above Côtil Point. A contact between the Mont Mado aplogranite and the coarser St Mary's granite is exposed here, and is marked by about 1 m of granite rich in mafic minerals. The aplogranite has been traced southwards to Handois, where it was once exploited as china-stone (see pp.106–107). Handois Quarry has now been dammed and is used for water storage.

Belle Hougue igneous complex

The Belle Hougue complex forms the headland of that name (Figure 3). It consists mainly of altered diorite, veined and brecciated by granite. The dioritic and granitic rocks are mineralogically like those at Ronez and Sorel points. Similar rocks are exposed on Les Sambues reef [663 565] to the north-east, and neither the submarine extent of the mass, nor its relationship with the north-west granite, is known. On Belle Hougue Point itself there is a large inclusion of Jersey Shale Formation [6553 5637] within the diorites. On the south side of the mass the Les Rouaux Fault has brought the plutonic rocks against the St Saviour's Andesite Formation on Belle Hougue Point and east of Les Rouaux, whereas in the small bay of Les Rouaux [657 563] they abut against the Jersey Shale Formation (see p.78). The fault is probably of considerable magnitude, for specimens collected from the complex all show the effects of cataclasis. The weathered and altered appearance is due in the main to cataclasis and accompanying mineralogical changes. The diorites contain ragged green amphibole, altered plagioclase and abundant epidote. The granitic rocks are traversed by fissures and the constituent minerals are broken and comminuted, so much so that in places the rocks have a mortar structure. The rocks vary both in the proportion of potassic feldspar to plagioclase and in the amount of quartz present. Most are adamellites but local deficiency in quartz makes some of the rocks monzonitic. Others, however, are rich in potassic feldspar and quartz, and have a granophyric texture.

It seems likely that the Belle Hougue mass is related to the north-west granite and has been separated from it by faulting.

South-west granite

The south-west granite was emplaced into the Jersey Shale Formation. Radiometric age determinations have given a Rb:Sr whole-rock isochron age of 553 ± 12 Ma (recalculated from 565 ± 15 Ma of Adams, 1976; Bishop and others, 1975), showing that it is older than most of the other Jersey granites. From the coast near La Carrière [5636 4948] eastwards almost to Pont Marquet [593 495] the contact between the granite and the Jersey Shale Formation is covered with blown sand, but thence to Belcroute Bay [6065 4800] it can be traced with some certainty, being markedly lobate, with tongues of granite extending into the Brioverian sediments which are baked against it. The degree of hornfelsing of the country rocks is everywhere slight and there is no marked aureole surrounding the south-west granite. The attitude of the sediments at St Aubin and elsewhere has not been greatly affected by the emplacement of the granite, though veins of granite up to 20 cm wide occur in the sediments near the contact. The form of the granite was studied by Henson (1956), who also noted that the granite becomes grey as the contact is approached, in contrast to the more usual pink. This band of grey, rather crushed and contaminated granite is about 100 m wide in Belcroute Bay, and contains large crystals of quartz and feldspar up to 1 cm in diameter in an even-grained matrix; sedimentary xenoliths decrease in size and abundance away from the contact. This grey porphyritic marginal facies is exposed at several places inland, in each case within a few metres of the contact.

The marginal facies apart, the south-west granite has three main types: a coarse-grained equigranular granite—the Corbière granite; a porphyritic granite with megacrysts of potassic feldspar set in a finer-grained matrix—the La Moye granite; and an aplogranite—the Beau Port granite. The three are arranged roughly concentrically.

The coarse-grained Corbière granite has the largest outcrop. It passes in a short distance into the La Moye porphyritic granite which forms a band some 200 m wide between the other two types. The groundmass of the La Moye granite is very similar to the Beau Port aplogranite which is exposed around St Brelade's Bay. The Corbière granite contains abundant blocky crystals of orthoclase perthite, together with plagioclase, quartz, mica and hornblende. The granite forms spectacular, murally jointed cliffs along much of the south-west coast of the island, and weathers to a pale red colour. Inclusions are rare and are most abundant at the granite margins. Biotite and hornblende are everywhere associated, and accessory zircon, apatite, iron oxides, fluorite and allanite occur.

The La Moye porphyritic granite is variable in texture, composition, and size and abundance of the megacrysts. It grades on the one hand into the Corbière coarse-grained granite, and on the other, into the Beau Port aplogranite where the megacrysts are fewer in number. Intrusive contacts between the various granite types are unknown and do not provide evidence on which to assign relative ages to them. In places, for example in the extreme south-west of Portelet Bay [5965 4670], the porphyritic granite has a banded appearance, caused by concentrations of biotite and opaque minerals. Such bands occur close to the aplogranite, and between them the granite is markedly porphyritic. The banding is a local phenomenon apparently produced as a result of interaction between the aplogranite and the porphyritic granite.

The Beau Port aplogranite differs from the other members of the complex in its texture and in its deep red weathering. It is uniformly even grained and, compared with the other granites, fine grained, though its grain-size averages 1 mm. The principal minerals are quartz, perthite, and plagioclase (oligoclase An_{12-16}), with small amounts of biotite, iron oxides and hornblende.

The La Moye granite was interpreted by Henson (1956) as Corbière granite modified by the emplacement of the later Beau Port aplogranite. However, A. M. Bland (1984) has shown that there are marked chemical differences between the Beau Port aplogranite on the one hand and the Corbière and La Moye granites on the other, as well as considerable differences in age.

The Corbière and La Moye granites were formerly quarried in several places, but this activity has now ceased.

South-east granite

The south-east granite is the most complex of the three principal plutonic masses. As in the north-west complex, the first rocks to have been emplaced were gabbros which were altered to diorites as a result of interaction with granitic material.

The Dicq granite forms an intrusive complex with diorite, which is well exposed on the intertidal reefs of La Grève d'Azette, and the emplacement of this granite was responsible for much of the alteration of gabbro to diorite. The Dicq granite contains megacrysts, usually up to 2 cm long, of pink to grey orthoclase perthite, commonly with a white rim of plagioclase, set in a coarse-grained matrix of quartz, potassic feldspar, plagioclase, biotite and hornblende. Xenoliths are abundant in this granite; some are clearly of diorite but others, usually rather rounded and about 10 cm long, are more fine grained; the latter commonly contain megacrysts of mantled potassic feldspar identical to those in the granite. Adams (1976) obtained an imprecise Rb-Sr isochron for the Dicq granite which gave a date of 570 Ma (recalculated from 580 Ma), comparable with that of granitic rocks in Guernsey.

The Longueville granite extends inland from Dicq Rock [659 477] to Longueville, and is generally similar to the Dicq granite but lacks the potassic feldspar megacrysts. It has a characteristic yellowish weathering and is composed of a coarse-grained aggregate of plagioclase, potassic feldspar, quartz, hornblende and mica. The content of quartz varies from place to place, so that parts of the mass are syenitic. The granite typically contains many basic inclusions of gabbroic or dioritic origin. No intrusive contact between the Dicq and Longueville granites is exposed, and the granites probably grade into one another by variation in the proportion of potassic feldspar megacrysts.

The La Rocque granite occupies a large area extending from Grouville Arsenal to Gorey, La Rocque and the intertidal reefs in St Clement's Bay. In Petit Portelet bay it is assumed to be faulted against volcanic rocks. On the foreshore near the slipway [7107 5019] west of Gorey Harbour and in a small quarry [7114 5030] in the Gorey area the granite has intruded the Jersey Shale Formation. Most of the mass is pink-weathering granite composed of potassic feldspar, usually perthitic, with plagioclase and abundant quartz, and with biotite and hornblende as the most common ferromagnesian silicates. Zircon, apatite and allanite are accessories. The mass is not of uniform texture, however; the grain-size increases eastwards from Le Hocq [685 467], and around La Rocque the granite contains large, squat, tabular crystals of orthoclase perthite, some of which have a thin mantle of plagioclase. The granite at Mont Orgueil Castle [716 503] has a more brownish tinge than elsewhere, and quartz is locally less abundant here than is normal. Furthermore, the granite contains a few inclusions of rhyolite, which refute Plymen's (1921) contention that a flow-banded dyke of rhyolite indicated that rhyolite had flowed over the exposed surface of the granite and into fissures within it. Aplite and pegmatite are present as patches throughout the mass, and pegmatites closely associated with aplites are exposed on the reefs near La Rocque [707 463].

The Fort Regent/Elizabeth Castle granophyre is a compact rock that weathers to a yellowish pink colour. It has intruded diorite at Elizabeth Castle, with the production of sharply bounded angular inclusions of diorite in granophyre.

Near the Hermitage [6392 4732], however, granophyre has penetrated the diorites as subhorizontal sheets which have crenulated margins that cannot be the result of simple dilation intrusion. There is also abundant evidence of reaction between host diorite and intruding granophyre, with the production of grey rocks intermediate in character between the two. These dioritic rocks extend eastwards of the Castle to the reefs at La Collette.

The granophyre is composed of tabular plagioclase, usually partially altered to sericite, and mantled by potassic feldspar which is graphically intergrown with quartz. Ferromagnesian minerals are rare and are mainly chlorite after biotite. The micrographic texture is well developed, with feathery branching intergrowths extending outwards from a common plane in the orthoclase that is coincident with a crystal face.

Unlike the other igneous complexes, there is clear field evidence in the south-east complex for two pulses of granite emplacement. The older comprises the Dicq porphyritic granite and the Longueville granite. The diorite/Dicq granite complex has been intruded by basic dykes with an ENE–WSW trend and about 1 to 2 m wide. In several places at La Grève d'Azette [e.g. 663 467] the diorite/Dicq granite/basic dyke complex has been further intruded by veins of a later, pink, even-grained granite, which is similar to the main mass of the La Rocque granite; these veins cut across some of the early basic dykes in the diorite/Dicq granite complex. Reaction has occurred between the diorite and the newer granite, but reaction between the basic dykes and the granite was minimal. Around Green Island and Le Nez Point it is difficult to separate the effects of the Dicq and La Rocque granites so far as the diorites are concerned. The late pegmatites, and the reaction that accompanied veining by recognisable La Rocque granite, are ascribed to the second pulse of granite emplacement, but much of the reaction could belong to either intrusive event.

A further phase of dyke intrusion followed the emplacement of the La Rocque granite and these later dykes have the same general trend as those truncated by this granite.

The concept that the south-east complex was composed of granites of different ages goes back to Plymen (1921) and Groves (1927; 1930); it derived from the observations that three granite 'masses' (the Elizabeth Castle/Fort Regent, Longueville and Gorey granites) were aligned, that they were more syenitic than the younger granites, and that they contained different suites of heavy minerals. Detailed mapping has cast doubt on the idea of linearity and, although it is true that the Longueville and Gorey granites are locally deficient in quartz, isotopic dating indicates that they are of different, not similar, age; moreover, experience has shown that heavy minerals are unreliable indicators of age.

The relationship of the granophyre to the granites is far from clear. Aplitic granite, but not granophyre, occurs close to the Dicq granite at and south of Havre des Pas bathing pool, but sharp contacts between the two are wanting. Adams's Rb:Sr whole-rock isochron for the south-east granite (509 ± 4 Ma, recalculated from 520 ± 4 Ma) included the Elizabeth Castle granophyre, and geologically it is grouped with the La Rocque granite rather than with the older Dicq granite. Henson (1956) compared the Elizabeth Castle granophyre mass with the Beau Port aplogranite, and superficially they have much in common, but pink aplogranite is associated with all the Jersey granites.

There remains an unresolved problem resulting from isotopic age determinations. Field evidence indicates that the St Saviour's Andesite Formation followed the Jersey Shale Formation and was succeeded by the outpouring of acid volcanic material. All were deformed together during the Cadomian orogeny. The Jersey Shale Formation is older than 553 Ma, and it was intruded by the Dicq/Longueville granite also, though actual exposures of the contact are not available in the alluvium- and loess-covered inland areas. If the age of about 570 Ma for the Dicq/Longueville granite is accepted, and on geological grounds it must be older than 509 Ma, then the emplacement of gabbro and diorite must predate this, and isotopic determinations in both Jersey and Guernsey suggest a date of about 580 Ma for this event. The simplest interpretation for the poorly exposed diorite at Les Augrès is that it is an intrusive mass which is younger than the andesites into which it was emplaced; the andesites, therefore, are presumably older than about 580 Ma. Duff (1978), however, obtained a Rb:Sr whole-rock isochron age of 522 ± 6 Ma (recalculated from 533 ± 6 Ma) for the andesites from West Mount [646 493], and this was interpreted as the age of extrusion. If this is so, then it is possible to reconcile this age with Adams's isotopic dates only if the 570 Ma age of the Dicq granite is discarded or if it is assumed that the andesite has been incorrectly correlated; further, it would be necessary to interpret the Les Augrès diorite as a mass of older diorite faulted against younger andesite. Bishop and Mourant (1979) have suggested a reconciliation by questioning the assumption that Duff's date represented the time of outpouring of the andesite, because the andesites at West Mount Quarry contain feldspathic patches and the quarry is only 100 m or so from the Elizabeth Castle granophyre which has given a 509 Ma age (recalculated from 520 Ma). It could be that Duff's isochron represents an event later than the solidification age of the andesites. An ^{40}Ar:^{39}Ar date of 477 ± 6 Ma (formerly 470 ± 6 Ma) has been obtained for toumaline separated from andesitic tuff at Maison de Haut [6760 4925] (Allen, 1972).

Minor igneous intrusions   6

Minor igneous intrusions abound in all the Channel Islands. In Jersey, most are dykes (in the generally accepted usage of the term for vertical or near-vertical intrusive sheets) but a few have a subhorizontal sill-like form. Petrographically the intrusions fall into three principal groups – basic, acid, and lamprophyric – each containing several varieties. Field relationships indicate that the intrusive history has been complex.

The dykes are particularly numerous in the south of the island. This main swarm has a general W – E trend in south-west Jersey, but on the eastern side of St Aubin's Bay, at Elizabeth Castle, the alignment is slightly north of east. At La Collette it takes up the WSW – ENE orientation that is characteristic of the swarm in south-east Jersey, where the dykes are concentrated within a belt about 2.7 km wide and well exposed in the intertidal reefs. These dykes are exposed sporadically inland and were intersected by boreholes at Queen's Valley; they reappear on the east coast between Gorey and Archirondel (for example, at Jeffrey's Leap, Figure 15), apparently having been displaced somewhat to the north.

Basic intrusions

Basic intrusions are more abundant than the other petrographic types. Most are dykes, ranging in thickness from 10 m or more to the thinnest veins, the average being about 1 m. Although most are simple intrusions, multiple and composite dykes also occur (Figure 16). Virtually all the composite dykes have basic margins and porphyritic microgranite (feldspar-quartz-porphyry) interiors (Plate 13), but in a few this central portion is of lamprophyre. Although the composite intrusions are usually considerably thicker than the simple dykes, the basic margins are of comparable thickness – usually less than 1 m – and represent only a small part of the total thickness of the dyke. Wherever the order of intrusion has been established, the basic margins appear to have been emplaced first, followed by the acid or lamprophyric interior.

There are several compositional and textural varieties of basic dyke. The two principal ones are: dolerites containing either fresh or relict pyroxene, with or without amphibole;

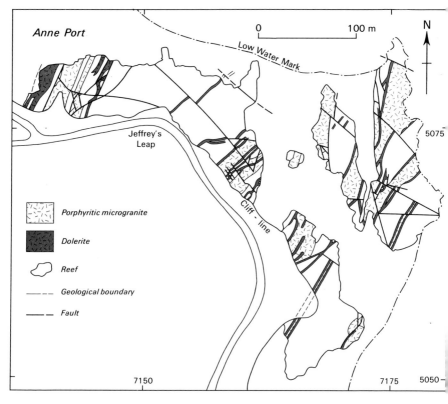

Anne Port

0   100 m

Low Water Mark

N

Jeffrey's Leap

Cliff - line

5075

Porphyritic microgranite

Dolerite

Reef

Geological boundary

Fault

7150

7175

5050

and basic intrusions showing various degrees of alteration both of feldspars and of coloured minerals. Aphyric, porphyritic, and ocellar varieties of each occur. Amphibole and chlorite are abundant in the altered dykes, which range from rocks in which original textures are preserved to those which are so altered as to resemble epidiorites. Brown amphibole also occurs in some of these altered dykes. The degree of alteration seems to be largely independent of the order of intrusion of the dykes (see pp.68–70). Nearly all the dykes in the Jersey Shale Formation and in the overlying volcanic rocks are intensely altered, and many are sheared. On the other hand, dykes belonging to the main swarm of the southeast plutonic complex are principally dolerites around St Helier and at La Collette, but are almost entirely metadolerites farther east from La Grève d'Azette to Pontac. The alteration would seem not to be the result of weathering, but to be due to some form of low-grade metamorphism which affected the rocks as a whole.

Near Green Island (Figure 17) a coarse-grained basic dyke [677 461] 10 m wide has a dioritic mineralogy, and closely resembles the dioritic rocks into which it was intruded. This dyke has chilled margins and felsic pockets containing euhedral, elongated, cored amphiboles like those of the

Figure 15 Sketch map showing dykes exposed in reefs at Jeffrey's Leap. Redrawn from Thomas, 1977, fig.10.1

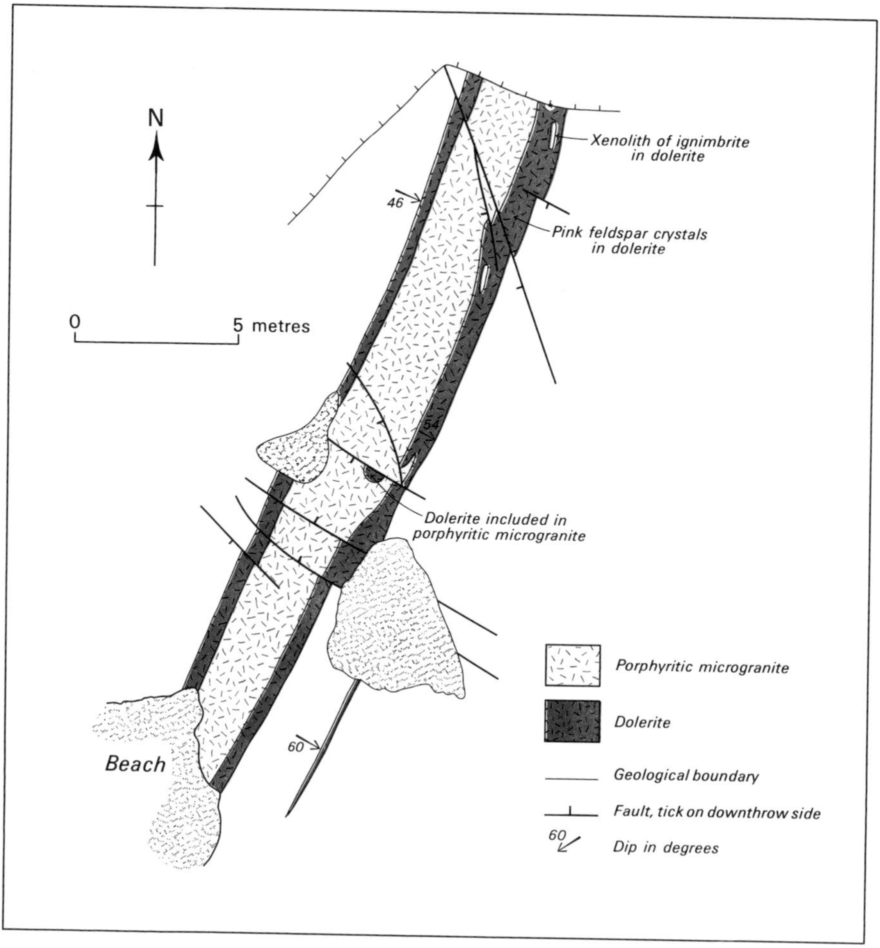

N
0
5 metres
Xenolith of ignimbrite in dolerite
46
Pink feldspar crystals in dolerite
54
Dolerite included in porphyritic microgranite
Beach
60
Porphyritic microgranite
Dolerite
Geological boundary
Fault, tick on downthrow side
60
Dip in degrees

Figure 16 Sketch plan of a composite dyke at Petit Portelet. Redrawn from Thomas, 1977, fig.10.4

hornblende-lamprophyres (see p.66). It is, however, distinct from them, for it intruded the camptonitic hornblende-lamprophyre.

Near Seymour Tower a few basic dykes in the La Rocque granite have the form of elongate pods separated from one another by granite. They are comparable to the synplutonic dykes described by Cobbing and Pitcher (1972) from Andean plutons, and indicate the overlapping of granite emplacement and dyke intrusion. These dykes have distinctive granoblastic fabrics, as do certain dark fine-grained inclusions in the diorites at Green Island. These inclusions differ chemically from the dykes of the main swarm and could be interpreted as early dykes, probably intruded into the original gabbro and subsequently broken up and altered. Small quartzose xenocrysts, probably indicative of contamination, are widespread in the dykes in the eastern part

of the main swarm. At Pontac [690 468] such dykes are cut by aplitic veins, again pointing to the near-contemporaneity of granite emplacement and dyke intrusion.

Eight basic sills, the thickest about 1 m, are exposed in the granite cliffs between Grosnez Point [549 567] and Le Pulec [549 547] on the north-west coast of the island. They dip gently south-east, so that the lowest is exposed in the north and the topmost in the south. Each individual sill is exposed along the strike for 100 m or more before thinning out. Three such sills – the greatest number in any one place – are exposed above one another near the former German radar tower at Les Landes [5450 5598], but the most spectacular is the sill which runs at the foot of Le Pinacle [5445 5545].

An extensively altered basic sill, about 0.5 m thick, is exposed on the western side of Portelet Bay [598 470] and at the foot of the cliffs surrounding the small bay immediately east of Pointe le Fret [596 468]. A differentiated composite sill, 8 to 10 m thick and dipping at 25° just east of north, is

Plate 13 A composite dyke at Jeffrey's Leap, Anne Port, has dark doleritic margins separated by pale porphyritic microgranite. (A13700)

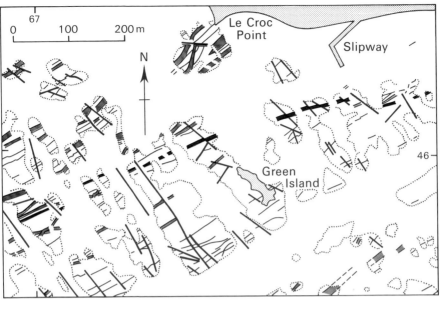

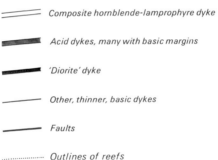

—————— Composite hornblende-lamprophyre dyke

▬▬▬▬ Acid dykes, many with basic margins

▬▬▬▬ 'Diorite' dyke

————— Other, thinner, basic dykes

————— Faults

·············· Outlines of reefs

Figure 17 Sketch map of the area around Green Island, showing the displacement of dykes by faults. Many small basic dykes have been omitted for clarity

exposed at La Collette [6525 4756] (Bishop, 1964b). This sill has chilled margins, and its main mass was probably emplaced as a basic magma which was permeated by acid fluids, for it contains potassic feldspar and brown amphibole, both concentrated in bands and distributed throughout the rock, giving it some of the characteristics of hornblende-lamprophyre. The sill contains several bands of porphyritic microgranite coplanar with the contact, the most prominent being about 0.75 m from the top.

Acid intrusions

Acid intrusions are fairly widespread in Jersey, but most are confined within the dyke swarm in the south-eastern part of the island. They are generally 4 to 10 m wide and, though some are simple intrusions, many are composite with relatively thin basic margins.

There are two principal types of acid intrusion, porphyritic microgranites (or feldspar-quartz-porphyries) and flow-banded rhyolites. The porphyritic microgranites are the more abundant; they are pink- to yellow-weathering rocks, usually containing abundant phenocrysts of quartz, potassic feldspar and plagioclase in varying proportions, set in a fine-grained matrix. The quartz phenocrysts show varying degrees of resorption, and most of the plagioclase is unzoned, twinned, albite-oligoclase. The groundmass may be granular, spherulitic, or micrographic. Spherulitic fringes of varying width mantle quartz and potassic feldspar phenocrysts, and some of the dykes have a completely micrographic matrix.

Three flow-banded rhyolite dykes have been recorded – at Jeffrey's Leap [715 508], south of Mont Orgueil Castle [7154 5022], and south of St Aubin [607 482] – but their relation to the other dykes is unknown.

Lamprophyres

There are two principal types of lamprophyre in Jersey, hornblende-lamprophyre and mica-lamprophyre. The hornblende-lamprophyre dykes are much less abundant than the mica-lamprophyres.

At South Hill [6504 4764] two leucocratic hornblende-lamprophyre dykes have been displaced by low-angle faults or 'shears' (Bishop, 1964a). Before land reclamation south of St Helier harbour covered the exposures, they could be seen to be cut by the La Collette sill [6497 4760], which was in turn intruded by basic dykes of the main swarm. Farther east, at Le Croc [673 462] and on the reefs to the WSW (Figure 17), a composite hornblende-lamprophyre has basic margins followed inwards by zones up to 1 m wide of porphyritic microgranite and a central mass of camptonite; this distinctive intrusion is demonstrably older than the main dyke swarm, having in places been cut by composite microgranite dykes, and it serves in assessing the amount of movement that resulted from later faulting (pp.80–81).

Several lamprophyres have intruded the Rozel Conglomerate. A composite dyke at Douet de la Mer [7018 5418] has a central mica-lamprophyre between altered basic margins, and another south of La Coupe Point [711 538] has basic margins and a wide hornblende-lamprophyre centre.

Most of the hornblende-lamprophyres are spessartites, with long green amphibole phenocrysts set in a matrix of small tabular plagiocase crystals. Some contain a little potassic feldspar, quartz and chlorite, together with accessory minerals. The hornblende-lamprophyre at Le Croc Point [673 462] has phenocrysts of brown amphibole and is camptonitic. Smith (1936b) described as a camptonite a rock from

Plate 14 A dyke of mica-lamprophyre in granite below Mont Orgueil Castle, Gorey. (Photograph by Dr A. C. Bishop)

near La Cotte à la Chèvre with phenocrysts of altered olivine, altered feldspar and fresh pyroxene, in a groundmass containing abundant small crystals of brown amphibole: it would now be grouped with alkaline olivine-basalts.

The mica-lamprophyres occur as rusty brown dykes, rarely more than 1 m wide (Plate 14), distributed throughout the island in all the major rock groups. They fall into two groups, one with a sharply defined, nearly N – S orientation, and the other scattered about a NW – SE direction. These intrusions have been described by Smith (1933, 1936a, b); they vary texturally, some having a nodular appearance and others containing ocelli, and also in points of petrographic detail, but they are the most clearly defined single group of minor intrusions.

The mica-lamprophyres, though somewhat variable, generally contain phenocrysts of biotite, many of which show colour-zoning in thin section, the margins of the crystals being more deeply coloured than the interiors. Phenocrysts of olivine (generally altered to serpentine, calcite and other minerals) and fresh pyroxene usually accompany the mica. These minerals are set in a groundmass of alkali feldspar. In most of the dykes the feldspar is fresh to turbid orthoclase, which occurs as large grains or as small radiating crystals, and justifies the name minette for these rocks. A few dykes (for example at Noirmont [607 464] and Creux Gabourel [571 563]) were stated by Smith (1936b) to contain sodic plagioclase and thus to be kersantites, but more recent work by Mr G. J. Lees indicates that they, too, may be minettes. Other lamprophyres were referred by Smith (1936b) to monchiquites because of their structureless groundmass.

Intrusive sequence

Intrusive histories have been deduced locally, but correlating these from place to place has proved difficult and it has not been possible to elucidate with certainty the overall chronographical sequence of intrusion for the Jersey dykes.

The basic dykes at Côtil Point [631 562] predate the intrusion of the Mont Mado aplogranite which, being part of the north-west granite, has a Rb:Sr isochron age of 480 ± 15 Ma (recalculated from 490 ± 15 Ma of Adams, 1976). It is not known how these dykes relate temporally to some of the early dykes in the Jersey Shale Formation.

The South Hill leucocratic hornblende-mica-lamprophyres are earlier than the La Collette sill which, in turn, has been intruded by basic dykes of the main swarm. It is doubtful whether the last are the 'older' basic dykes of La Grève d'Azette, for they and the sill have intruded dioritic rocks associated with the Fort Regent granophyre, one of the youngest of the south-east granites. Similarly, the camptonitic hornblende-lamprophyre at Le Croc (Figure 17) is cut by microgranite dykes which were themselves intruded by basic dykes.

Sills are rare in Jersey and all seem to be relatively early in the intrusive sequence. The basic sills between Grosnez Point and Le Pulec are cut by small aplite veins, and the sill just east of Pointe le Fret is cut by a N–S basic dyke.

At Noirmont and elsewhere it is clear that the main swarm of basic dykes was not emplaced as a single intrusive phase, because dykes can be seen to intersect. In south-east Jersey, dykes belonging to the main dyke swarm have intruded diorite and both the Dicq and La Rocque granites. On the intertidal reefs at La Grève d'Azette several basic dykes have intruded the Dicq granite but have themselves been truncated by aplogranite, presumably related to the La Rocque granite to the east. These older dykes have the same trend as those which have intruded both the Dicq granite and the aplogranite; because they are not petrographically distinct, it has proved impossible, in the absence of cross-cutting relationships, to distinguish older from younger dykes.

Throughout the island, basic dykes with a roughly N–S alignment cut across dykes with more nearly E–W trends, though a few N–S dykes are earlier than the E–W ones. The later N–S dykes are usually dolerites and are fresher than those they cut; a particular example, though more coarse grained than usual, is a dyke of olivine-gabbro 8 m wide at Wolf Caves [6347 5616] and its presumed continuation south of Côtil Point.

The main swarm dykes appear to have been emplaced during a period of crustal relaxation. The related extension of the crust varied in direction from N–S to NNW–SSE, and judged from the thickness of the dykes its amount compares with that associated with the Tertiary dyke swarms of Mull and Skye; also, its duration was probably relatively short, about 1 to 2 Ma, similar to the estimated period for the emplacement of the Tertiary dyke swarms of Scotland (Speight and others, 1982). Dykes were probably emplaced sporadically, the later ones cutting across the earlier under the control of local features such as joints. The 'microgranite' composite dykes show spectactularly such local discordances, especially at La Grève d'Azette, but these are probably basic dykes which have tapped a source of acid magma at depth, so that special significance should not necessarily be attached to these features. Such distinctive dykes, however, show clearly – at Le Croc Point, for example – that emplacement and faulting were closely associated. The host rock to the dyke swarm, the La Rocque granite, has been dated at 509 ± 4 Ma (recalculated from 520 ± 4 Ma of Adams, 1976), so placing an older age limit on the dyke swarm. The evidence for a close temporal relationship between granite emplacement and dyke intrusion suggests an Ordovician age for the dyke swarm.

The basic, hornblende-lamprophyre and mica-lamprophyre dykes that have intruded the Rozel Conglomerate Formation (see p.40) indicate that dyke emplacement went on well beyond the cooling period of the plutons. Adams (1976) obtained a K:Ar date of 427 ± 13 Ma (recalculated as 434 ± 13 Ma) from hornblende from the hornblende-lamprophyre south of La Coupe Point, and though this age is best regarded as tentative it sets a minimum date for the deposition of the Rozel Conglomerate. Most indications from the occurrences of mica-lamprophyres in the Armorican–Hercynian belt of Europe point to a late Carboniferous–early Permian date for their emplacement (Lees, 1982).

The N–S dykes are fewer than the dykes of the main swarm and they were emplaced when crustal extension was directed roughly E–W; they are also chemically distinct from the dykes of the main swarm, being more alkaline in character. Thick N–S basic dykes occur in Brittany, where they have intruded the Cap Fréhel Sandstone of Lower Palaeozoic, possibly Cambro-Ordovician, age: alkaline basalt dykes also occur in Guernsey (the albite-dolerites). The Jersey dykes may be related to either or both of these dyke suites. Isotopic ages are not available for the acid, basic, and mica-lamprophyre dykes of Jersey, but Adams (1976) obtained a Variscan K:Ar date of 317 ± 9 Ma (recalculated as 324 ± 9 Ma) from a dolerite dyke at Roselle Point, Alderney. He also obtained a K:Ar date of 296 ± 8 Ma (now 303 ± 8 Ma) for a mica-lamprophyre from Petit Port, Guernsey, indicating a Variscan intrusion age.

Duff (1980) studied the palaeomagnetism of the Jersey dykes. He separated the dykes into three groups, A, B, and C, all of which had trends and thicknesses characteristic of the main dyke swarm. The palaeomagnetic ages suggested for them – Carboniferous, Silurian to Middle Devonian, late Precambrian to early Cambrian, respectively – though supporting an extended intrusion history, are nevertheless surprising. None of the demonstrably late N–S dykes appears to have been investigated. It is hard to envisage how the Jersey lamprophyres (presumably mica-lamprophyres and included by Duff in Group B), which field relationships show to be the youngest of all the dykes, should have given a Silurian to Middle Devonian pole position, older than the alleged Carboniferous pole of the Group A dykes. At present it is difficult to reconcile the palaeomagnetic conclusions with the field relationships or with the few isotopic dates available.

Genesis

The dykes of Jersey – with the exception of the later N–S ones, the mica-lamprophyres, and certain early dykes intruding the Jersey Shale Formation – make up one coherent suite of rocks. This coherence, reflected in their geochemical composition, transcends differences in dyke direction, mineral composition, and relative age of intrusion. The chemical characters of the dykes have a distinctly bimodal distribution, in which the basic rocks (dolerites, metadolerites), characterised by relatively low content of SiO_2, and high MgO, FeO and CaO, are distinguished from the acid rocks ('microgranite'), with high SiO_2, Na_2O and K_2O, and low MgO, FeO and CaO.

The basic rocks are chemically subalkaline and show typical calc-alkaline patterns for the distribution of major elements and the trace elements Y and Zr. However, calc-alkaline rock suites do not normally show bimodal distribution of elements because most rocks in such suites are of intermediate composition (andesitic). Intermediate rocks among the Jersey dykes are confined to the hornblende-lamprophyres of La Collette and Le Croc Point, and are present in such a small quantity relative to the basic and acid rocks that it is difficult to explain the chemistry of the rocks in terms of crystal fractionation from an initial basic magma; similarly the trace-element distribution in the acid rocks would be difficult to obtain by such a process.

In view of the close temporal relationship between the emplacement of granite and of the dyke swarm, a more plausible explanation would involve the mixing of the incoming basic magma with the already emplaced granite. In this model, the material forming the acid dykes would represent not the end product of fractional crystallisation of basic magma, but granite magma remobilised from the host pluton at a fairly high level in the crust. The few intermediate rocks would be the product of mixing (by hybridisation or contamination) of the basic and acid end members. Such mixing would appear to have occurred only at an early stage in the history of the swarm to produce the leucocratic hornblende-lamprophyre of La Collette and the camptonite of Le Croc. Later mixing would seem to have taken place only locally, though acid and basic magmas continued to co-exist.

Geological structure 7

Folds in the Jersey Shale Formation

The Jersey Shale Formation is best exposed at the northern and southern ends of St Ouen's Bay, and at St Aubin. In these and other areas the tectonic evolution of the formation has been studied recently by Squire (1974) and Helm (1983, 1984). Squire believed that the earliest folds in the Jersey Shale Formation, representing the main (Viducastian) phase of the Cadomian orogeny, were originally open in form, with horizontal E–W-trending axes, but that they were subsequently tilted to give steep plunges. He described a second set of folds with north-westerly axial trends and generally gentle plunges towards the south-east; these folds were open and asymmetrical, with average wavelength of about 700 m, axial planes inclined south-westward, and a poorly developed cleavage; he considered that this folding was also Cadomian and penecontemporaneous with the emplacement of the granite. Squire remarked that an E–W zone, dominated by gentle dips and folds with NW–SE trends, extends from Mont à la Brune [582 506] to Tesson Mill [617 508] and beyond.

Helm (1983) worked on the Jersey Shale Formation exposed in the intertidal reefs in St Ouen's Bay, and later (1984) extended his investigation to the remainder of Jersey. He recognised two main phases of deformation, D_1 and D_2 (Figure 18), but found no evidence for the early phase of N–S compression noted by Squire. The early (D_1) folds are of two types, either singly plunging (Plate 15) or doubly plunging (periclinal). The axes of the periclinal folds trend from WNW–ESE, through N–S, to NE–SW; the periclines are generally asymmetrical, open to close, gently plunging, and mainly with a westerly vergence, but some are upright or have a slight easterly vergence. Helm concluded that the D_1 periclines probably represent early-formed non-cylindrical buckles initiated by irregularities in the bedding owing to channel infills. The singly plunging folds are open to close, and some are isoclinal; they are commonly asymmetrical, generally with a Z-shaped profile and a mainly gentle to moderate south-easterly or south-westerly plunge; most have a dextral vergence. Many of the singly plunging D_1 folds have a relatively strong, spaced, axial-planar cleavage, but no mesoscopic fabric is usually associated with the periclines. Helm suggested that the singly plunging D_1

folds are parasitic on the eastern limb of an inferred major D_1 anticline, whose axial trace must lie some distance to the west of the outcrops in the intertidal reefs. He considered that his D_1 folds might be equivalent to Squire's second set of folds. He also showed that in the 'domes and basins' area off La Crabière described by Casimir (1934) most of the structures are singly plunging D_1 folds.

Although overall the dip of the bedding in the Jersey Shale Formation is eastward, there is a strip of westward-dipping beds running N–S through St Peter [596 516]. This local reversal is due to a major D_1 fold pair, the St Peter Syncline and a complementary anticline about 1 km to the east.

The original approximately N–S axial trend of the D_1 structures was modified by later N–S (D_2) compression; this produced major folds, a non-penetrative axial-planar fabric (S_2), and a system of conjugate shear faults. Helm's D_2 folds are well exposed at the northern end of St Ouen's Bay. To the west of Le Pulec the strike of the bedding is NW–SE, but it changes to N–S and then NE–SW farther south, thus outlining a major anticline (the St Ouen Anticline), plunging eastwards and having a wavelength greater that 3 km and a minimum amplitude of 1.5 km. Helm noted that open to close parasitic folds, with wavelengths of 50 to 100 m, are present on each flank of the St Ouen Anticline; the associated cleavage is vertical to steeply inclined, and has a mean strike direction of 080°. The D_2 folding has changed the orientation of the axial traces of the D_1 folds, and on a smaller scale has resulted in cross-cutting cleavages and curvilinear D_1 fold hinges. Helm also recorded that the D_1 and D_2 folds have been overprinted by a system of late radial fractures attributed to vertical stress generated by the uprise of the basaltic magma which filled some of the fractures. In

Plate 15 Jersey Shale Formation beds of Association III in intertidal reefs opposite the Slip de L'Ouest at the northern end of St Ouen's Bay showing large-scale singly plunging D_1 folds with a parasitic fold pair. (Photograph by Dr D. G. Helm)

addition he suggested that the sporadic occurrence of closely spaced N–S joints might indicate that there had been a fourth deformation, but he detected no associated folds, although there was a similar fabric in the adjacent north-west granite.

Figure 18 Sketch map showing the main structural features of Jersey. Based on Helm, 1984, fig.4

Folds in the Jersey Volcanic Group

Benefiting from the lithological distinctions that he had made in the sequence of volcanic rocks, Thomas (1977) was able to recognise folds of several sizes and three orientations, namely E–W, N–S, and NE–SW. In the St Saviour's Andesite Formation at West Park, the St Helier Syncline plunges to the south-west (Figure 18) and has been intruded by granophyre with cross-cutting relationships. This fold is separated by an anticlinal area from the large north-eastward-plunging Trinity Syncline in north-east Jersey. The Trinity Syncline affects both the Jersey Shale Formation and the volcanic rocks (Teilhard de Chardin, 1920; Mourant, 1933), but is modified by many smaller folds trending either E–W or N–S. The interference of the smaller folds has given rise to domes and basins, and some of the domes have brought andesite to the surface (commonly beneath loess) within the main Bonne Nuit Ignimbrite outcrop, for example north-west of Le Grès [669 524] and at intervals in the valley to the south.

Helm (1984) confirmed most of the structures in the volcanic rocks described by Thomas; he listed D_1 folds, including a possible fold pair near Frémont Point and a syncline in Bonne Nuit Bay (both trending NNW–SSE), the E–W folds just north of Archirondel Tower (see p.35), and a refolded and faulted syncline in Vallée des Vaux. D_2 folds include anticlinal flexures that give rise to the inliers of the Jersey Shale Formation at Le Bourg and Gorey (the St Saviour Anticline; see p.6). Helm also observed three previously unrecorded cleavages in the volcanic rocks, the general NW–SE, NE–SW and N–S trends of which suggest that they are axial-planar to similarly oriented folds mapped by him and Thomas, and described above.

Folds in the Rozel Conglomerate Formation

The palaeotopographical surface at the base of the Rozel Conglomerate has been folded into an open syncline (D_3) with a sinuous axial trace trending WNW–ESE. Local departure from the NNW–SSE strike of bedding suggests that the main syncline may have been warped about NE–SW-trending axes (D_4), though large-scale cross-bedding makes confirmation difficult; these flexures are associated with closely spaced joints and shear zones.

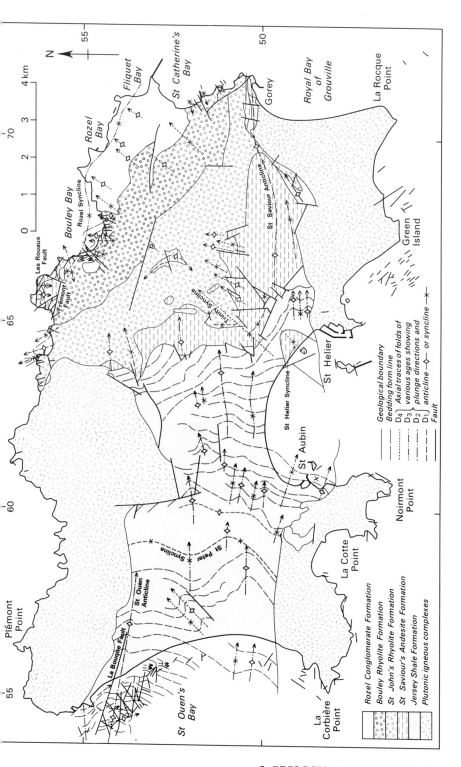

Helm (1984) and Richardson (1984) have shown that the Rozel Conglomerate has a strong, generally NW–SE-trending pressure-solution fabric associated with shearing and rotation of pebbles (Plate 16). Most of the pebbles show tectonic pitting, and the granitoid pebbles are extremely distorted and commonly interdigitate with adjacent metasedimentary clasts. In places, for example below La Tête des Hougues [679 546], mudrock layers have a slaty cleavage and contain reduction spots in the form of strain ellipsoids which indicate a flattening of about 38 per cent.

Faults

The main faults that are known to occur in Jersey are shown in Figure 18. Faults can be traced most readily on the foreshore, especially where rocks of contrasting lithology are juxtaposed; elsewhere important faults may have gone undetected. The dykes are of particular value in demonstrating the effects of faulting. The distribution of the dykes in the intertidal reefs of south-west and south-east Jersey (Figure 17) has revealed a fault pattern that cannot be traced inland owing to poor exposure; the formational boundaries drawn on the map are the best approximation from the available outcrop evidence. In addition, many dykes occupy fault fissures.

The principal faults strike WNW–ESE, but many other trends are represented. Squire (1974) noted that most of the faults are nearly vertical, and that wrench faults predominate. Some of the faults predate the emplacement of the granites; for example, the E–W sinistral wrench fault north of Mont Mado Quarry is truncated by the Mont Mado granite [6366 5588]. Some faults are also earlier than the Rozel Conglomerate, as shown by the fact that the outlier of conglomerates at Pierre de la Fételle has not been affected by the Frémont Fault (below); however the Rozel Conglomerate is displaced by a later sinuous fault that follows the E–W valley to the south of Rozel Manor [698 533].

The La Bouque Fault trends WNW–ESE, roughly parallel to the nearby margin of the north-west (St Mary's) granite, and limits the southward extent of thermal spotting in the shales. Folds and quartz veins in the rocks adjacent to the fault indicate a dextral movement that Squire (1974) put at more than 1 km; however, the amount of the displacement is uncertain, owing to the lack of marker horizons. Of the faults in the Jersey Shale Formation exposed in the intertidal reefs to the south, several share the strike of the La Bouque Fault, but other trends have been traced, many of them between NNE–SSW and NE–SW.

The fault in St Peter's Valley trends a little west of north at its northern end but is nearer SSE–NNW at its southern

1 cm

Plate 16
Deformed pebbles
of granite and
Jersey Shale
Formation in
Rozel
Conglomerate
from Rozel Bay.
(Photograph by
Dr D. G. Helm)

end. The fault dextrally displaces the margin of the north-west granite by about 1.3 km. It cannot be traced farther with certainty, but it is possible that the fault exposed at Crabbé [596 556] is its northward continuation. A fault of NNE–SSW trend, separating the north-west (Mont Mado) granite from the Jersey Shale Formation, was noted by Mourant (1933) in the trench for the Handois Reservoir dam [6322 5374]; its sinistral displacement was put at 400 m by Squire (1974).

The Frémont Fault (Figures 3 and 7), called the L'Homme Mort Fault by Mourant (1933) and other authors, is a broad zone of sheared rocks or cataclasite that separates the Jersey Shale Formation from the Frémont Ignimbrite at Frémont Point (Plate 17). Drag folds in the shales indicate dextral shear (Thomas, 1977). The fault is again found at L'Homme Mort, in the south-east corner of Giffard Bay (Plate 18), and it can be traced to the west side of Bouley Bay, south of Vicard Point. Dr A. E. Mourant has suggested (personal communication) that the fault continues beneath the Rozel Conglomerate to near Rozel Manor, where it may have brought up an inlier of basalt (see p.23), although for this to be so the strike must change to NW–SE. The Frémont Fault was thought by Mourant (1933) to be a normal fault with a downthrow of 6000 ft (1830 m) to the south, but Casimir and Henson (1955) reduced this estimate to 2000 ft (610 m), whereas Squire (1974) suggested a dextral displacement of 1 km and a downthrow to the north-east of a few hundred metres. Thomas (1977) disputed these figures and preferred a lateral movement of nearly 4 km, with little vertical throw.

The Les Rouaux Fault trends generally E – W just south of Belle Hougue Point. It brings diorite (containing an enclave of Jersey Shale Formation) and granite of Belle Hougue type against Jersey Shale Formation and St Saviour's Andesite Formation (Plate 19; p.53). Squire (1974) recorded that it is a vertical fault with a downthrow to the south-east and an apparent dextral strike-slip component of about 300 m. Between this fault and the Frémont Fault, and also along the coast westward to La Saline [630 561], many smaller faults have been recorded. Similarly west and south of Bouley Bay, and along the coast between Archirondel and Anne Port, many minor faults of varied trends have been traced by distinguishing offsets of the members that make up the Bouley Rhyolite Formation. A fault largely concealed by loess is believed to course WNW – ESE to the south of Ville à l'Evêque [654 540] and Les Câteaux [670 533].

Plate 17 Frémont Point, St John. Jersey Shale Formation near sea level on the left is separated by the Frémont Fault from ignimbrites of the St John's Rhyolite Formation that make up the remainder of the headland. (A13669)

Plate 18 Shear zone of the Frémont Fault at L'Homme Mort, Giffard Bay. Fragments of Jersey Shale Formation, andesite and rhyolite are embedded in a fine-grained matrix. (A13704)

On the northern side of St Helier the Clos de Paradis Fault is exposed in Wellington Road [6618 4927], where the Jersey Shale Formation is in contact with andesite; the fault zone is about 3 m wide and is inclined at about 84° north-eastward. Mourant (1933) noted a further exposure of the fault in foundation trenches at Le Clos de Paradis housing estate, and traced the fault towards the top of Queen's Road. Squire (1974) considered this fault had a dextral displacement of about 1 km and an apparent downthrow to the north-east, but Thomas (1977) commented that a downthrow of a few hundred metres on a normal fault would also account for the observed disposition of outcrops. The fault has itself been displaced by lesser faults with trends between N–S and NE–SW. Again, to the north, several faults trending between NNW–SSE and NNE–SSW have affected the boundary between St Saviour's Andesite and Bonne Nuit Ignimbrite.

At the south-west side of the Victoria Marine Lake [642 488] a fault aligned NNW–SSE has dextrally displaced the axial trace of the St Helier Syncline. North and south of Elizabeth Castle, faults trending respectively about E–W and ENE–WSW have similarly moved the granophyre/diorite contact.

The hornblende-mica-lamprophyres at South Hill have been displaced by mylonite-filled faults which dip gently westwards. These faults have been cut by basic dykes of the main swarm, indicating that the movements took place before the major period of dyke emplacement.

Many of the Jersey dykes are subvertical – those on the east coast dip steeply south-east – and so it follows that the displacement of their outcrops must have been due to transcurrent faults. Most of the faults cut the dykes at about right angles, and though both dextral and sinistral displacements have been noted, the general movement was

Plate 19 Les Rouaux Fault east of Belle Hougue Point, Trinity. A deep gulley has been eroded along the fault plane between granite and diorite (on the left) and Jersey Shale Formation and members of the St Saviour's Andesite Formation (on the right). (A13703)

in a dextral sense, as can be judged by the displacement of the camptonitic hornblende-lamprophyre dyke and the 10 m-wide dioritic dyke south-west of Le Croc (Figure 17); the horizontal displacement of the dioritic dyke amounts to 300 m over a strike length of 1500 m. These faults also displaced the later N–S basic dykes, and the maximum stress was directed slightly west of north.

Near Le Hocq, however, the granite boundary is assumed to have been displaced sinistrally for a considerable distance, though this movement could have occurred before the emplacement of the dykes which seem not to be offset in this way. As noted above (p.61), however, the dyke swarm as a whole has apparently been displaced to the north between St Clement's Bay and its reappearance on the east coast northward from Gorey, although this need not necessarily have been the result of faulting.

Tectonic history

Helm (1984) provided a synthesis of the tectonic history of Jersey, which briefly is as follows. During the Cadomian orogeny E–W compression gave rise to the early D_1 folds, which had an approximately N–S orientation in both the Jersey Shale Formation and the volcanic rocks. D_2 folding resulted from N–S compression, which refolded the D_1 folds about E–W axes and produced a sinuosity in the N–S structural trend. (Helm suggested that the D_2 compression might be related in some way to the intrusion of at least part of the granite complexes.) After the Cadomian folding and granite intrusion the area was uplifted and eroded, and the products of erosion accumulated in hollows to form the Rozel Conglomerate. D_3 compression, directed NE–SW, gave rise to the NW–SE-trending Rozel Syncline. A further phase (D_4) yielded NE–SW-trending folds, in particular the major Trinity and St Helier synclines.

Thomas (1977, citing Key, 1974) speculated that the association of a thick sequence of acid volcanic rocks with granites might be indicative of a caldera collapse. He demonstrated the concept by turning the geological map of eastern Jersey anticlockwise through 90°, when the distribution of outcrops might represent an oblique section through the supposed caldera. However, having taken all the evidence into account, he concluded (personal communication) that, while it is possible that the ignimbrites may have been associated with a caldera, no trace of it can now be identified, and the present arrangement of volcanic and granitic rocks is fortuitous.

Quaternary deposits 8

The deposition of the Rozel Conglomerate was followed by a long period which has left no sedimentary record in Jersey. The major elements of the present outlines of the Channel Islands resulted from faulting, probably in Mesozoic times. Eocene (Lutetian) marine limestones beneath the sea to the north, west and east of Jersey are down-faulted against the older rocks of the island, and it is likely that intermittent uplift of the island as a block bounded by submarine faults continued from the late Mesozoic to the late Cenozoic. The planation surfaces that are prominent in the higher parts of the island probably originated during pauses in the uplift; their age is uncertain, but they may have been formed in the Neogene rather than the early Pleistocene, on the evidence of the low altitude of the Plio-Pleistocene deposits of the southern Cotentin peninsula on the French mainland about 35 km east of Jersey (Lautridou, 1982).

The youngest rocks in the island are of Quaternary age and include sediments of both the Pleistocene and the Holocene (Flandrian). They reflect the changing climates and sea levels from the Middle Pleistocene, when the earliest of these sediments was laid down.

As is commonly so, the age relationships of many of these deposits are obscure, and strict superposition of sedimentary units generally cannot be seen; thus the suggested sequence rests partly on geomorphological considerations rather than on the much more satisfactory criteria used for the hard rocks. Nonetheless the following succession, with estimated maximum thicknesses, has been recognised:

Holocene	Blown sand	20 m	Flandrian
	Peat and alluvium	10 m	Flandrian
Pleistocene	Loess	5 m	?Devensian
	Head	30 m	Devensian
	St Peter's Sand (older blown sand)	5 m	?Ipswichian
	8 m raised beach	3 m	?Ipswichian/Eemian
	Head	2 m +	?pre-Ipswichian
	18 m raised beach	2 m	?pre-Ipswichian
	30 m raised beach	2 m	?pre-Ipswichian

Except for the loess, which covers the upland plateau of the island, and the valley-side head deposits, these drift formations occur largely in coastal areas (Figure 19).

Pleistocene

Raised beaches The gravels of the 30 m and 18 m raised beaches are assumed to be pre-Ipswichian. Neither beach has been firmly dated, although ages in the upper part of the Middle Pleistocene and in the lower part of the Upper Pleistocene seem probable.

The 30 m beach is known from only two localities: South Hill, St Helier [6510 4770], and St Clement [6869 4738]. Together these show a height range between about 25 m and 37 m above mean sea level (Keen, 1978b). Only at South Hill can the sediments of this beach be examined, and there they consist of a medium gravel of the local Fort Regent granophyre and quartz pebbles.

The 18 m beach is present in caves on the north-west coast of Jersey and on fragmentary benches below cliffs which rise to the main plateau of the island, as at Jubilee Hill (Mont du Jubilé), St Ouen's Bay [5750 5145] (Renouf, 1969), St Helier (Renouf and Bishop, 1971), and at St Clement. In the caves it either occupies a bench cut into the solid rock or forms a false roof, as it does around Grand Becquet [5750 5628], where the beach is a cemented cobble gravel of local rock types. The best exposure is at the former terminus of the Jersey Eastern Railway at Snow Hill, St Helier [6543 4838](Plate 20), where medium gravel, composed almost entirely of local granite but with a few flint clasts (probably derived from offshore outcrops of Chalk), fills a gulley on the east side of the cutting. Both fine and coarse gravels are present at Jubilee Hill, and flints accompany the local rock types that make up the greater part of the deposit.

Recent work on the raised beaches of the south coast of England and South Wales, using the amino-acid racemisation technique (Davies, 1983; Davies and Keen, 1985), suggests that at Portland, Torquay, Swallowcliffe (Avon), and in Gower, beaches reaching similar heights to the 18 m beach of Jersey have ages of around 200 000 ± 20 000 years.

The 8 m raised beach on Jersey occurs widely around the coast, but its best exposures are in the south-west from La Corbière Point [553 481] to St Aubin [606 482], in the north in Bonne Nuit Bay and Giffard Bay, and in the north-east from Rozel Bay [6965 5465] to Anne Port [7135 5110]. The beach material ranges in size from cobbles, as at Bouley Bay [6688 5495], to fine gravel and sand, as in the section north of Belcroute slip [6069 4810]. It consists largely of local rock types, but it also contains small quantities (less than 5 per cent) of flints, quartzite like that from the Grès Armoricain

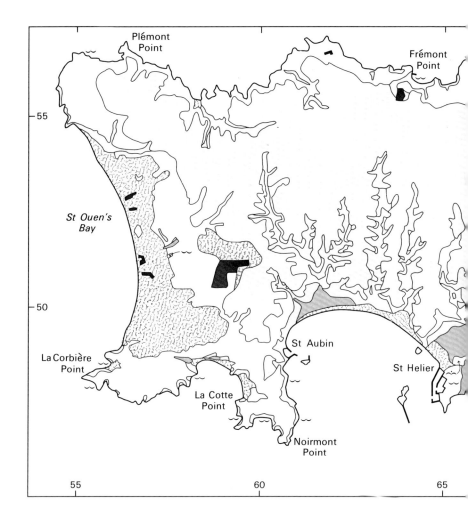

Figure 19
Geological sketch
map showing drift
deposits of Jersey

of Normandy and Brittany, and a few pebbles of Devonian limestone (Dunlop, 1911; Keen, 1975). It is very unusual to find flints or other exotic constituents of the raised beach larger than 5 cm in the long axis. The beach ranges in height from 11 m above mean sea level south of La Cotte Point [5928 4753] down to 3 m above mean sea level, as exemplified by the beach exposure [608 465] north of Noirmont Point, where cemented gravel occurs within the modern beach (Keen, 1978b).

In the south-west of the island the 8 m raised beach gravels normally rest on head, rather than on a smoothed surface of solid rock. Such a sequence can be seen clearly at Portelet Bay [6015 4710] and Belcroute Bay [6070 4825] (Figure 20), where the beach is separated from the rock platform by up to 2 m of head, presumably pre-Ipswichian in age. At Portelet the coarse boulder beach, with clasts up to 40 cm in the long

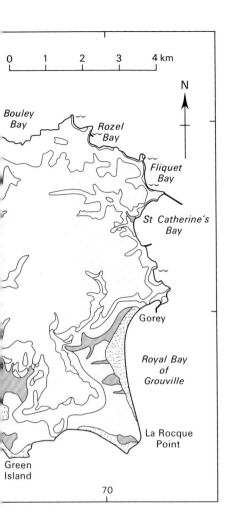

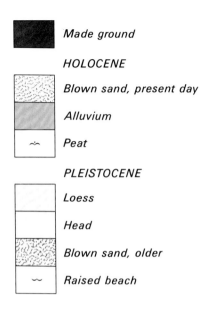

0 1 2 3 4 km

N

Bouley Bay
Rozel Bay
Fliquet Bay
St Catherine's Bay
Gorey
Royal Bay of Grouville
La Rocque Point
Green Island

70

Made ground

HOLOCENE

Blown sand, present day

Alluvium

Peat

PLEISTOCENE

Loess

Head

Blown sand, older

Raised beach

axis, is composed entirely of local rock types. The finer gravel in Belcroute Bay contains small amounts of flint and quartzite. At both Belcroute and Portelet the beach gravel is overlain by about 2 m of sand; at Portelet the sand is poorly stratified and passes up into head, but at Belcroute it is well bedded, with individual layers up to 2 cm thick. It is uncertain whether these sands are water-laid and thus part of the beach proper, or small accumulations of blown sand formed during the retreat of the sea from its highest level.

Along the whole of the north coast of Jersey the wave-cut notch associated with the 8 m raised beach is the main feature defining the cliff base. Gravel of the 8 m raised beach occurs in Bonne Nuit Bay [6422 5588], at the Belle Hougue cave [6560 5640] and north of the jetty at Bouley Bay [6688 5495]. At the last locality a beach of rhyolite cobbles overlies 2 m of head as at Portelet. In Bonne Nuit Bay, beach gravel is visible at the base of the head, and 200 m east of the jetty [6424 5589] a beach gravel strongly cemented by ferruginous minerals occurs within the modern beach in the same way as that at Noirmont Point. Only in the cave known as Belle Hougue I does the 8 m beach contain any molluscan remains; here nine species of Mollusca are preserved in stalagmite, and indicate a sea temperature at the time of formation of the beach up to 3°C above that of the present English Channel (Zeuner, 1940). This suggests than the 8 m beach was deposited during the Ipswichian/Eemian interglacial, which is known to have been warmer than the present (Shackleton and Opdyke, 1973). This age is supported by amino-acid ratios from shells of *Patella vulgata* (Linné) from the raised beach at Belle Hougue, which suggest correlation with other 8 m beaches of probable Ipswichian age in England and Wales (Davies, 1983), and by U-series dates from the stalagmite cement of this beach, which have yielded an age of 121 000 $^{+14\,000}_{-12\,000}$ years, closely similar to that generally agreed for the Ipswichian/Eemian in Europe (Keen and others, 1981).

In the north-east the 8 m beach generally rests on a rock platform and consists of up to 1 m of gravel overlain by head. The clearest exposures occur around Belval and north of Rozel [6962 5460] but small patches of beach occur discontinuously in the base of the head along the whole coastal section from Rozel to Gorey. The beaches here are mostly of fine gravel or sand containing few cobbles. As in the southwest, most (more that 90 per cent) of the constituents are of local origin, but flints and quartzites also occur.

St Peter's Sand (older blown sand)   The St Peter's Sand, which occupies a hilltop position around the eastern part of Jersey Airport, is possibly of similar age to the 8 m raised beach. There are no sections in the sand but temporary exposures and auger holes indicate a maximum thickness of 5 m. The deposit is a well-sorted, structureless and iron-cemented sand of quartz and feldspar. Its stratigraphic relations are obscure but it appeared to underlie the loess in a road cutting [5937 5052] in the southernmost part of its outcrop. Thus, if the loess is Devensian, a pre-Devensian age for the sand is possible. Because it is well sorted, an origin as blown sand seems probable (Keen, 1975) and this sand sheet is perhaps an Ipswichian counterpart of the Flandrian sand in the same area.

Head   Head occurs at the foot of cliffs along the south-west, north, and north-east coasts of Jersey. It also forms undissected cones and fans beneath the blown sand at St Ouen's, St Aubin's, and St Clement's bays, and at the Royal Bay of Grouville, and it mantles valley sides inland,

Plate 20 Deposits
of the 18 m raised
beach overlying
Fort Regent
granophyre at
Snow Hill, St
Helier. (A13707)

where it has been derived mainly from the loess of the
plateau surface. As a whole the head is unsorted and composed of the reworked weathered remnants of the local solid
rocks.

In the south-west the greatest thickness of head is about
15 m, and the major sections are in Beau Port [579 481]
(Plate 21), in the bays south of La Cotte Point [5945 4715],
in Portelet Bay, and from Noirmont Point northward to St
Aubin. Immediately south of St Aubin's Harbour the head
has mainly formed by weathering of the Jersey Shale Formation; the largest clast size is about 30 cm, but the fragments
are generally around 2 cm across; these coarser components
rest in a matrix of silt derived from the loess and the
weathered shale, and the silt is bedded, the beds being up to
5 cm thick with depositional dips of up to 30°. Elsewhere in
the south-west the head is formed entirely of frost-shattered,
angular fragments of the local granite; the maximum size of
the clasts is controlled by the spacing of the joints in the
granite, the largest being about 2 m across, and the larger
fragments are set in a gravelly matrix composed of granules
of quartz and feldspar; except along the east-facing cliffs
north of Noirmont Point, little loess occurs in this head.

In the north, the greatest thicknesses of head in Jersey lie
against the high cliffs of volcanic rocks in Bonne Nuit and
Bouley bays. The head thickness is related to the height of

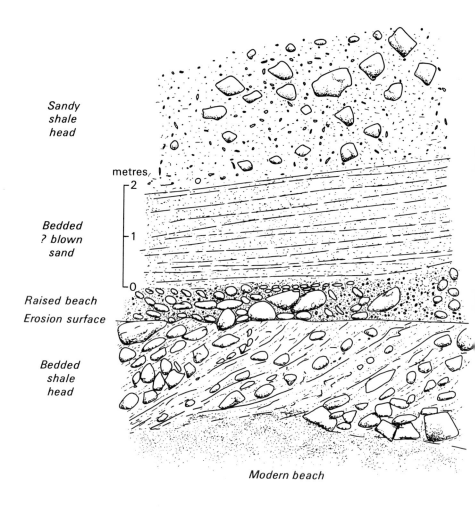

Sandy
shale
head

metres
- 2
- 1
- 0

Bedded
? blown
sand

Raised beach
Erosion surface

Bedded
shale
head

Modern beach

the cliffs, in the ratio of cliff height to head thickness of 4 or 5 to 1. The maximum head thickness is about 30 m, although the coastal exposures show rather less than 20 m. Below the outcrops of well-jointed rhyolite and andesite the head is very coarse, with clasts up to 2 m across in a matrix of finer fragments of volcanic rocks; there is little loess in these head deposits. At the base of the sections in the middle of Bonne Nuit Bay [6450 5585] and near the jetty [6415 5590] the head is composed of fine laminae of grey silt, perhaps indicating deposition in pools on the former foreshore left as the sea level fell at the end of an interglacial stage.

On the north-east coast, head occurs from Rozel eastwards to La Coupe Point [710 540] and southwards to Petit Portelet. In the northern part of this section the head is finer grained on the outcrop of the Rozel Conglomerate than it is elsewhere; the largest clasts measure 5 cm along their

Figure 20 Sketch section of the drift deposits at Belcroute, St Aubin's Bay

longest axis and are set in a matrix of loess. In the southern part, the clasts are composed of volcanic rocks and are up to 1 cm across in a matrix of loess; the best sections are at Petit Portelet, at Jeffrey's Leap [7160 5070], and north of Anne Port [7140 5130]. North of Belval Cove the loess content is sufficiently high to form discrete bands within the head; these are particularly well seen on the south side of the cove below La Coupe [7103 5385] as beds of loessic head up to 40 cm thick; such beds are distinguishable from primary loess by their deeper orange colour and by the few granules of Rozel Conglomerate that they contain.

Inland, on the relatively gently sloping valley sides, the head is only 2 to 3 m thick. It is generally a fine-grained deposit formed from reworked loess with incorporated granules of the local solid formations. This is best seen in the side valleys to the main south-trending valleys in the centre of the island, for example in road sections in St Peter's Valley and St Lawrence Valley; a road section at La Fontaine St Martin [6235 5240] shows 3 m of head on the upper part of the valley side.

The head is largely without faunal remains, but at La Cotte de St Brelade it has yielded a range of mammalian material, including *Mammuthus primigenius* Blumenbach, *Coelodonta antiquitatis* Blumenbach, *Rangifer tarandus* Linné, and Rodentia, as well as Aves, and tools and skeletal remains attributed to *Homo sapiens neanderthalensis* King. In coastal exposures south of St Aubin, palaeolithic implements have occasionally been found (Mourant, 1935; Keen, 1978a), providing evidence of man's presence during head deposition. At Fliquet Bay [7115 5350] the basal part of the head contains about 60 cm of stony, silty peat, which floors a gulley cut into the Rozel Conglomerate. This peat has yielded a fauna comprising 38 taxa of Coleoptera and a pollen spectrum, both of which suggest that the environment was sub-arctic immediately before the head was deposited, although they give no indication of age (Coope and others, 1980). A radiocarbon date for wood fragments from the base of the organic deposits gave an age greater than 25 550 years (Birm-955), so that the head in the north-eastern part of Jersey dates from before the last main cold episode of the Devensian (Weichselian) glaciation of northern Europe. Organic muds, up to 30 cm thick, occupying shallow depressions on the rock platform cut into the Jersey Shale Formation at La Vau Varin [6071 4834], south of St Aubin's Harbour, have also yielded pollen and Coleoptera (Coope and others, 1985); these confirm the indications of the Fliquet deposits that the climate immediately prior to the deposition of the head was sub-arctic. At Ecalgrain (Manche), on the Normandy coast, peat in a similar stratigraphic position to that at Fliquet has yielded an age greater than 45 000 years (Shotton and Williams, 1971), thus placing the overlying head in the early Devensian. The Jersey head is underlain by the ?Ipswichian/Eemian 8 m raised beach and overlain by Flandrian blown sand, so that the head as a whole is probably Devensian in age.

Loess　The loess is an orange-yellow to pale brown sediment in which 80 per cent of the particles are of silt grade (Plate 22). It is composed largely of quartz and feldspar, but it is calcareous in patches. At Green Island (La Motte) [6740 4596] decalcification textures (*limon à doublets*) can be seen. The lower part of this profile contains redeposited calcium carbonate in the form of discontinuous concretions (*lössmänn-chen*) of loess cemented by $CaCO_3$; it is around and within these concretions that a sparse molluscan fauna of *Pupilla muscorum* (Linné), *Oxyloma pfeifferi* (Rossmässler) and *Trichia hispida* (Linné) occurs. Mollusca of these species have also been found in calcareous loess, or very loessic head derived from it, at Portelet [599 472], St Aubin [606 485] and Belval Cove [7098 5275] (Keen, 1982). Bovine bones have been found in the loess at Pontac.

The main areas of loess in Jersey are in the centre and east of the island. The greatest thicknesses of about 5 m occur on the coastal plain in St Clement and on the plateau around La Hougue Bie [6830 5036]; it is about 3 m thick in St

Plate 22 Loess
resting on diorite
and overlain by
blown sand and
Neolithic soil at
Green Island, St
Clement.
(A13714)

Lawrence and it thins westwards until, in St Peter and St Ouen, the cover is discontinuous. In places, as between Grouville church [6930 4854] and Gorey Village [705 500], the loess occurs on lee slopes facing east and north-east.

The well-sorted nature and lack of structure of the loess are consistent with an aeolian origin, and the sediment was perhaps derived from the floor of the English Channel exposed during a period of low sea level.

The age of the loess is not known, but the fact that it is overlain by blown sand – for example, at Green Island – indicates that it predates the Flandrian. The fauna, though sparse, suggests that accumulation took place in cool or cold climatic conditions, and thus points to an origin during one or more of the Pleistocene glacial episodes. Several periods of loess deposition are known to have occurred in Normandy (Lautridou, 1973, 1982) and, although no evidence of this has been found in Jersey, by analogy with Normandy it is possible that the lower levels of the Jersey loess are of considerable antiquity in the Pleistocene, especially in the east where the deposit is thickest.

Holocene

Peat and alluvium Peat and alluvium are so closely related as to form one sedimentary sequence for the purpose of description. They occur as narrow deposits along most of the river valleys but the main occurrences are at the mouths of the major valleys from St Aubin to St Helier and Georgetown, and at Queen's Valley south-west of Gorey; smaller amounts of alluvium and peat are located at St Ouen's and St Brelade's bays, and at Grouville. The seaward edges of the alluvium and peat are commonly overlain by blown sand.

The thickest known deposits of peat and alluvium occur in the St Helier basin and at Grouville Marsh, where up to 8.5 m are present; the sediments are largely organic silts and muds, but peat layers up to 2 m thick also occur. The coastal peats at St Brelade's and St Ouen's bays are less than 1 m in thickness.

The sediments are predominantly freshwater in character, but most coastal areas show silt layers and pollen evidence, confirming that at least two episodes of marine transgression occurred during deposition (Birnie and others, in preparation).

The oldest peat and alluvium is found in the base of the fill in the main valleys; at Quetivel Mill, St Peter [6135 5120], the base of 2.65 m of organic mud has yielded a boreal forest pollen assemblage dated by radiocarbon to 9670 ± 70 years before the present time (SRR-2639). The coastal peats are younger, with radiocarbon dates of 4030 ± 60 (SRR-2634) from the base of the peat at L'Etacq [5530 5435], and 3150 ± 90 (SRR-2637) at L'Ouaisné [5951 4761]; however, south of L'Ouzière [5650 5150], in St Ouen's Bay, the upper surface of 70 cm of peat is dated to 3984 ± 50 (SRR-437), so deposition here began earlier, perhaps prior to 5000 years before the present. At Don Street, St Helier [6539 4871], a peat 1.69 m below street level yielded a date of 2660 ± 70 (SRR-2638); this indicates continuing organic deposition at a time when peat accumulation at the coast had been stopped by blown sand deposition, as shown by a date of 3470 ± 60 (SRR-2633) from the top of the L'Etacq deposit.

Blown sand The Holocene blown sand adjoins St Ouen's Bay in the west of Jersey, Grève de Lecq in the north, St Brelade's, St Aubin's and St Clement's bays in the south, and the Royal Bay of Grouville in the east. All these sand accumulations, except that in the west, are restricted to a coastal strip less than 500 m in width. Sand thicknesses of up to 27 m have been recorded at St Ouen's Bay, but the average thickness is probably 15 m or less.

At St Ouen's Bay (Plate 23) the blown sand is typically a quartz-feldspar sand with low shell content. The sand is largely structureless, except for a few gently dipping planar beds revealed in 'blow-outs' in Les Blanches Banques [5750 4986]. Buried soil profiles appear as humic lines within the sand in this section, and are marked by the occurrence of the

Plate 23 Dunes of blown sand at Quennevais, St Brelade. (A13660)

land gastropods *Cernuella virgata* da Costa, *Cochlicella acuta* (Müller) and *Pupilla muscorum* (Linné), and a few flint implements and Neolithic pottery shards.

At the northern end of St Ouen's Bay the sand extends 1.5 km inland from the coastline and in the south it has spread 3 km inland to Pont Marquet [593 495]; in both areas it reaches to the surface of the plateau. At the southern end of the bay the sand is up to 15 m thick where it is banked against the Middle Pleistocene fossil cliff, but its general thickness is less, around 10 m under the coastal plain and only 2 to 3 m on the plateau (Keen, 1981).

The radiocarbon date (SRR-439) from the surface of the peat that underlies the sand at L'Ouzière has shown that sand-blowing began in St Ouen's Bay after 3980 years ago and has continued intermittently to the present day. Although much of the sand has been stabilised by Machair vegetation, dune formation is active in the south of the bay north of La Carrière [5645 4959], where the wall built by German forces in 1940–45 has been buried to its top.

Blown sand occurs in patches at Beau Port, and at St Brelade's and L'Ouaisné bays where it overlies head. At St Helier blown sand rests on peat and alluvium below the Old Jail site [6470 4885]. As at St Ouen's Bay, the blown sand along the south coast is a quartz-feldspar sand with few structures, although at St Clement's Bay the sand contains some shell debris derived from the shell-sand of the beach. On the evidence of the radiocarbon date from the peat below the blown sand at L'Ouaisné [595 476] sand-blowing began after 3150 years ago, and continued through the Iron Age when the site at Ville-ès-Nouaux [6345 4985] was buried (Hawkes, 1938), to the present time.

The blown sand at St Clement's Bay connects with that on the east coast by way of a small deposit overlying the loess inland of La Rocque Point. The sand in the east is similar to that elsewhere in the island in being of limited landward extent and largely structureless, although with a greater shell content than in St Ouen's Bay. To the east of the mouth of Queen's Valley [7018 4939] the blown sand overlies the alluvial fill of Grouville Marsh, proving an age relationship like that in other parts of the island.

The deposition of the alluvium/peat sequence and of the blown sand were both related to the Flandrian rise in sea level. The alluvium and peat were deposited in meres or lagoons ponded behind the beach bar of the rising sea, and the blown sand accumulated as the Flandrian transgression reached its maximum and beaches became established in about their present positions.

Geophysical field surveys 9

Gravity survey

A gravity survey of Jersey was carried out by Briden and others (1982) as part of a systematic coverage of the Channel Islands. A total of 206 observations was made using a La Coste-Romberg gravity meter, giving a station density greater than 1 per square kilometre and generally confining the standard error in field readings to ± 0.02 mGal (1 mGal = 10^{-5} m/s²). This compares with the original study by Day (1959) at a station density of less than 1 per 3 km² and field reading accuracy of ± 0.1 mGal. Gravity measurements were referred to the National Gravity Reference Net 1973 datum at Jersey Airport (Masson Smith and others, 1974), where observed g was determined as 980 991.720 ± 0.009 mGal, because the original St Helier base station occupied by Day (1955) no longer existed. Data were reduced by means of the 1967 International Gravity Formula, and terrain corrections (out to 21.9 km about each station) and Bouguer corrections were made for a nominal density of 2.67 g/cm³. Elevations were referred to mean sea level, and wherever possible benchmarks or other levels for which an accuracy of ±0.01 ft was claimed were utilised; elsewhere, spot heights quoted to the nearest 0.1 ft or 0.5 ft by Hunting Surveys Ltd were used. For the less accurate spot heights, the uncertainty in elevation resulted in an error of about ± 0.05 mGal in the Free Air and Bouguer anomalies, in addition to other measurement errors.

The Bouguer anomaly map (Figure 21) was computer contoured using the SACM routine refined by Z. K. Dabek of the Applied Geophysics Unit of the Institute of Geological Sciences (now the British Geological Survey); this map is available from BGS as an overlay to the 1:25 000 geological map, and the data have been incorporated into the Bouguer Gravity Anomaly Map, Guernsey Sheet (Institute of Geological Sciences, 1979). The Bouguer anomaly field (Figure 21) is dominated by a large, circular positive anomaly, which is centred several kilometres east of St Helier and affects the Bouguer contours over the eastern two-thirds of Jersey. A circular symmetrical positive anomaly has therefore been subtracted from the Bouguer data to facilitate recognition of smaller-scale 'residual' and longer-wavelength 'regional' anomalies. Because this idealised procedure is undoubtedly an oversimplification, using it to define the shape

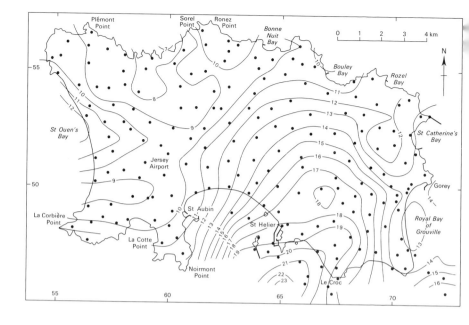

Figure 21
Bouguer anomaly map of Jersey. Contours are drawn at 1 mGal intervals. Station locations are indicated. Redrawn from Briden, Clark and Fairhead, 1982, fig.5a

of the positive anomaly, and hence the geometry of its causative body, gives rise to residual anomalies that are spurious in the sense that they merely reflect departures by the real anomaly from the circular model. The existence of a regional gravity gradient across the Channel Islands was demonstrated by Briden and others (1982); however, across the island of Jersey the regional component of the gravity field can be considered uniform at 9 mGal, except possibly on the extreme west of the island. This constant has been removed from the Bouguer data before interpretation (Figure 23).

The centre of the model main circular anomaly was placed in the valley north of Bagot [6655 4875] by visual fit to the Bouguer contours (Figure 21). Bouguer anomalies were plotted at a function of radial distance from the chosen centre (Figure 22a), and a polynomial was fitted to those data which were unbiased by local features. The adoption of the background field as a circularly symmetrical positive anomaly as shown was justified by the good grouping of data in the plot and the lack of overall trend to the residuals.

The 9 mGal positive anomaly (Figure 22b) requires the existence of a subsurface body with density at least as great as that of the St Saviour's Andesite Formation, since the anomaly does not extend into the outcrop area of the andesites. The density contrasts required to produce realistic models of the causative body suggest a minimum density of 2.90 g/cm^3, and hence imply dioritic or more basic rocks. Although the positive anomaly does not match the surface geology in detail, it does overlie the various dioritic bodies within the south-east granite. These are believed to be metasomatised gabbro (see the section on gabbro and diorite, Chapter 5) and thus the most likely cause of the gravity anomaly seems to be a major dioritic or gabbro body at depth.

A quantitative attempt has been made to interpret the anomaly, using the three-dimensional modelling program of Cordell (1970). Two possible (best-fit) model types are shown in Figure 22b, representing circularly symmetrical lenses with diameters of some 12 km. The upper model has a flat top constrained at a depth of 1 km which results in a maximum thickness for the causative body of 1 km for density contrast of 0.3 g/cm^3 (or 1.5 km for 0.2 g/cm^3). The lower model has the centre of the causative body fixed at 1.75 km depth and gives a similar thickness to that found for the upper model. Portions of the model extending beyond the eastern coastline are outside the survey area and have no data to support them. An alternative interpretation, though less plausible, is that the anomaly is wholly or partly due to a thick andesite sheet at small depth within the Jersey Shale Formation of south-east and central Jersey, similar in shape to the postulated gabbro body.

The residual Bouguer anomaly map (Figure 23) has been derived by subtracting the background anomaly deduced in Figure 22. The residual anomalies within the area that contained the circular positive anomaly are generally of low amplitude (less than 1 mGal) and many do not correlate with the known surface geology. The larger-amplitude anomalies are considered first.

Along the south coast of the island the gravity anomaly increases southwards by up to 7 mGal. The gravity gradient and the maximum anomaly attained on the edge of the survey area together imply the existence of a basic body at least 750 m thick, overlying the proposed gabbroic lens associated with the circular anomaly. This interpretation is consistent with submarine occurrences of gabbro south of Jersey (Lefort, 1975). In the Rozel area, a 3 mGal anomaly 'high' parallels the coast and has no closure on the seaward side. This anomaly is also visible on marine gravity maps (Bacon, 1975). Modelling by extrapolation from surface geology results in improbable thicknesses of several kilometres for the Rozel Conglomerate, even using increased density contrasts. The most likely cause of this anomaly is an offshore basic intrusion or possibly a thickening of the Brioverian supracrustal sequence comprising the Jersey Shale Formation and the Jersey Volcanic Group (see below).

The remaining large-amplitude anomalies relate directly to the outcrop distribution within Jersey, and have been analysed using a two-dimensional modelling program. The

Figure 22 Derivation (a) and interpretation (b) of the principal positive Bouguer anomaly. Sections of models computed for density contrasts of 0.2 and 0.3 g/cm³. Based on Briden, Clark and Fairhead, 1982, fig.6

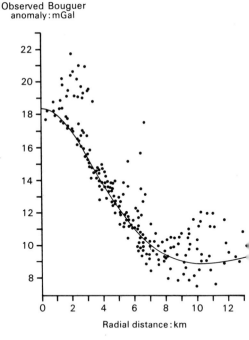

Observed Bouguer anomaly : mGal

Radial distance : km

a

calculations assume bodies of infinite strike, and thus yield slight underestimates of true size if the assumed density contrasts are correct. Three N – S profiles across western Jersey (Figure 23) have been interpreted in Figure 24 to suggest that the north-west and south-west granites are continuous beneath the Jersey Shale Formation, at depths ranging from 1800 m for profile A to 600 m for profile C. This compares with a depth of 300 m estimated by Day (1959) using a larger density contrast. All three profiles show the gravity 'high' associated with the south coast of Jersey, and profile C shows a positive anomaly ascribed to dioritised gabbro within the north-west granite at Sorel Point, where a thickness of at least 400 m of diorite is possible. The northern granite/ metasediment contact in profile C could not be fitted by any geologically viable model which is consistent with the adjacent profiles; this may be due to loss of two-dimensionality in the modelling, since the granite is flanked to its east and west by higher-density metasediments.

Elsewhere over the Brioverian sequence the residual anomaly field is almost uniform, despite significant density

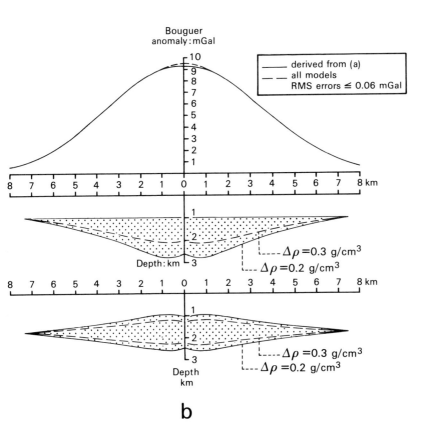

Bouguer anomaly : mGal

derived from (a)
all models
RMS errors ≤ 0.06 mGal

$\Delta\rho$ =0.3 g/cm^3
$\Delta\rho$ =0.2 g/cm^3

Depth: km

$\Delta\rho$ =0.3 g/cm^3
$\Delta\rho$ =0.2 g/cm^3

Depth km

b

Table 1 Rock densities

	Locality	No. of specimens	Density g/cm^3
Rozel Conglomerate	La Tête des Hougues	10	2.71 ± 0.01
Bonne Nuit Ignimbrite (rhyolite)	Bonne Nuit Bay	11	2.69 ± 0.01
St Saviour's Andesite (andesite)	St Helier	11	2.84 ± 0.02
Jersey Shale Formation (metasediments)	St Ouen's Bay	10	2.71 ± 0.01
Diorite	Ronez Quarry, Sorel Point	34	2.94 ± 0.02
Granite	Ronez Quarry, Sorel Point; Le Croc, La Grève d'Azette	45	2.63 ± 0.02

The densities in the table have been taken as representative of the major units of Jersey for gravity modelling and interpretation. The values given are the mean and standard deviation of laboratory measurements on saturated samples. Those from quarries are fresh; the remainder, from coastal exposures, are less so.

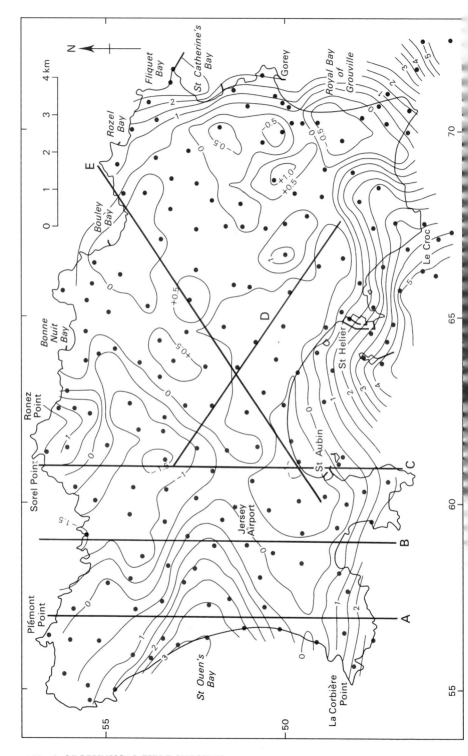

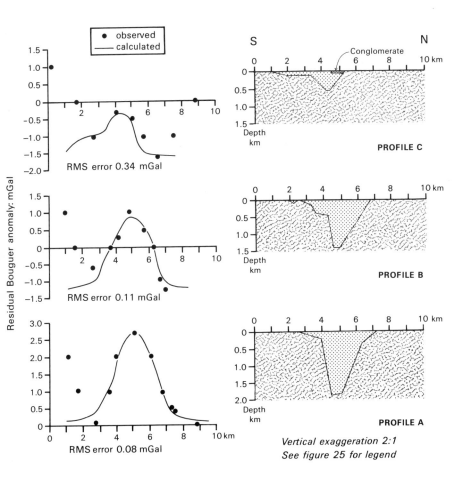

Figure 24 North–south profiles across western Jersey. Locations are shown in Figure 23. Models computed assuming a density contrast of 0.08 g/cm³; the inferred maximum depth of the metasediments varies by 200 m for a change in density contrast of 0.01 g/cm³. Based partly on Briden, Clark and Fairhead, 1982, fig.9

Figure 23
Residual Bouguer anomaly map of Jersey. Contours are drawn at 0.5 mGal intervals by the same routine as those in Figure 21. Locations of profiles in Figures 24 and 25 are shown. Redrawn from Briden, Clark and Fairhead, 1982, fig.8

contrasts between the rhyolites, andesites, and metasediments (Table 1). Gravity gradients across the contacts of the Brioverian with the three granites are low. It is therefore inferred that the Brioverian is a raft about 250 m thick, with little gravitationally resolvable structure, overlying rocks of granite density, implying that granites are continuous at no great depth beneath much of the island (Figure 25).

Those residual anomalies which do not relate to the known geology are short-wavelength features and therefore have a shallow origin. They are considered to reflect the topography

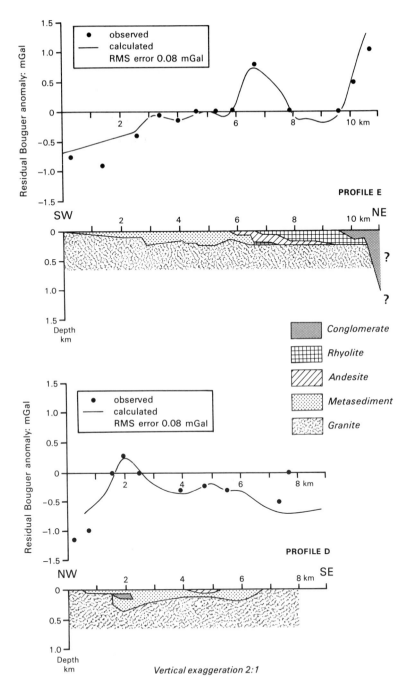

Figure 25 Profiles drawn diagonally across Jersey. Locations are shown in Figure 23. The granites are shown to be continuous below most of the island

of the granite/Brioverian interface, rather than any heterogeneities within the granite or small-scale structures within the underlying gabbroic lens. The depth of this boundary would be greater if a subsurface andesite body were evoked as contributing to the cause of the main positive anomaly (p.97).

Magnetic survey

Apart from three unpublished German measurements of declination, magnetic field observations are not known to have been made on Jersey prior to a reconnaissance total field survey made in 1976 over the eastern third of the island by Leeds University workers (Figure 26). Field readings were taken to ± 1 nanotesla (nT), and were reduced to total magnetic field anomalies using the 1975 International Geomagnetic Reference Field (IGRF); no regional trends were removed from the data. Briden and others (1982) remarked that a large magnetic anomaly trending N–S across the island has its western side coincident with the westernmost outcrop of the St Saviour's Andesite; this anomaly is broader than the outcrop, suggesting that the andesites extend eastwards near the surface beneath the rhyolite sequence, consistent with the interpretation of the residual Bouguer anomaly field (Figure 25). Other significant features in Figure 26 are the anomaly along the northern contact of the south-east granite, and the small magnetic signature of the diorite bodies within this granite. The former anomaly is most pronounced where the contact rocks are andesites, and these may be the source of the anomalous magnetisation. However, growth of magnetite and hematite is a common feature of metasomatic effects at granite contacts, and this would be an alternative explanation, although it has not been investigated specifically.

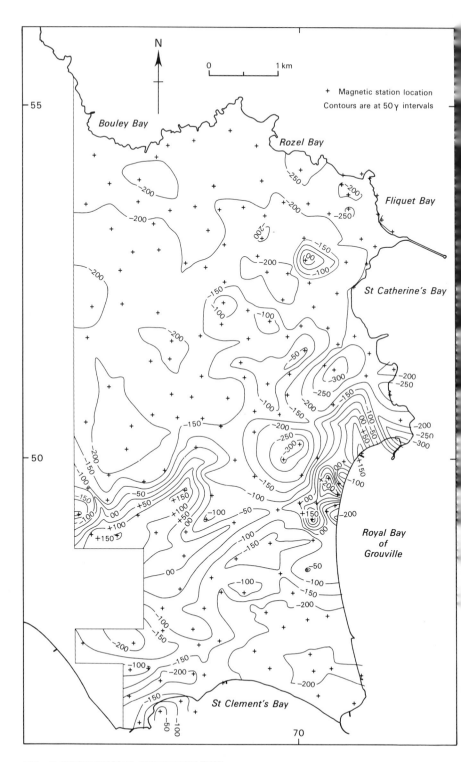

Figure 26 Total
field magnetic
anomaly map of
eastern Jersey.
Contours are
drawn at 50 nT
(= γ) intervals

Economic geology 10

In the past the population of Jersey was much smaller than it is at present and the needs of the islanders were met from the island's own resources, though they have always been dependent on the outside world for such basic commodities as metals and cement. The present economy depends principally on tourism, agriculture, and finance, and does not rely on the exploitation of natural resources, though water supply is of critical importance.

Quarrying

In the earliest times the basic need for shelter was met by caves. Later, the first builders – the constructors of the Neolithic dolmens – used wattle and daub for their houses, but slabs of rock for their tombs. Some of these rocks were of considerable size, and many were transported several miles (Mourant, 1977); thus granite from Mont Mado was taken to La Hougue Bie and other sites in Neolithic times. Also, the builders seem to have preferred certain rock types to others; for instance, Fort Regent granophyre is common in dolmens. In due course the abundant rocks suitable for building in virtually every part of the island were put to use; granite and diorite were particularly favoured, and builders simply took material from the beaches or removed blocks from reefs or outcrops using rudimentary tools.

As the need for more substantial buildings for housing, commerce and defence increased, so quarries were opened, though Rybot (1947) commented that quarries as they are known today did not come into being until the 18th century and even then were worked in a primitive manner. Nevertheless there is evidence that the Mont de la Ville, on which Fort Regent now stands, was probably quarried both by the dolmen builders and in the Middle Ages, before it was extensively worked during the late 18th century. By the 19th century the demand for rock for major building and construction works had so increased that new quarries were being started. Between 1847 and 1856 the large quarry at St Catherine [712 531] supplied material for the construction of the nearby breakwater. Though this is built mainly of Rozel Conglomerate, many of the granite coping stones were imported, probably from south-west England. This impressive structure was designed to be the northern arm of a large harbour, and a smaller quarry in ignimbrite at Archirondel [7092 5180] was opened to supply rock for the incomplete southern arm, which however, like its northern counterpart, also consists mainly of Rozel Conglomerate. The conglomerate for the southern arm was transported by rail from a quarry in the valley that runs eastwards to the sea from north-east of St Martin's Church.

Although rock was transported within the island from quarry to site, it made sense to use material that was locally available, and rock was quarried on the foreshore at low tide. There are evidences of such workings in many places on the south coast of the island. Jersey granite is difficult to work and to carve, and much granite was therefore imported, especially from the Chausey Islands. Chausey granite lacks the close jointing of the Jersey rocks and is easier to work, and it is thought that the technique for the accurate splitting of stone was introduced by Chausey workmen (Rybot, 1926). The use of Chausey granite for elaborate carvings in churches, for the carved troughs and wheels of crushers for cider making, as coigns and lintels of houses, and in Jersey fireplaces, extended over many years and has been documented by Stevens (1965, 1977). Rock for building was also imported from Les Minquiers and was used in the construction of Fort Regent. A considerable amount of Les Ecréhous gneiss – more than appears to have come from Les Minquiers – was imported from the 17th century (and possibly earlier) until at least 1811 and was used extensively in buildings, principally around Rozel (Mourant, 1956). At this time quarrying on Les Ecréhous was encouraged by the Admiralty in the hope of removing the islets altogether, as they were thought to be protecting French warships from British naval guns.

Towards the end of the 19th century the general demand for rock was such that its export – chiefly to Great Britain – became an economic proposition. Quarries at or near the coast were best situated for this purpose. Noury (1886) mentioned the commencement of quarrying at La Houle, and the resulting Ronez Quarry [617 568] is the largest working quarry in Jersey (Plate 24); great quantities of rock have been exported, chiefly for use as setts and kerbstones, and as road metal. The Mont Mado Quarry [637 556] at St John was much older and yielded perhaps the most handsome of all the Jersey monumental and building stones, a slightly pinkish grey aplogranite. Plees (1817, p.310) commented 'The stone from this spot is preferred to any other in the island. Many of the most ancient houses are built with it. It works readily, and a flat surface is easily obtained.' The quarry was filled some years ago.

Farther south, at Handois [6322 5386], the same facies of the north-west granite was quarried as a china-stone, which

Plate 24 Ronez Quarry, St John. (Photograph by Dr A. C. Bishop)

was reported to fuse more easily than that from Cornwall and was hence used for pottery glazes. The china-stone was discovered in about 1866 and it was quarried, crushed on site, and then shipped for use in the pottery industry (Furnival, 1904). Molybdenite that was found in this quarry was set aside and sold separately, probably to mineral dealers but possibly also as ore.

Rock was not quarried only for masonry: crushed rock – especially rhyolite from Anne Port – and even masons' chippings were used as road metal and for macadam. In places, for example at Longueville, Grouville Arsenal and La Moye, the rocks are so deeply decomposed that they were worked directly as gravel for use on the roads and some was even exported to Britain for making gravel paths.

The German occupation of Jersey from 1940 to 1945 saw a resurgence in quarrying to supply aggregate for concrete gun emplacements, defensive positions and anti-tank walls. Not only was the output of existing quarries increased and directed to this end, but old quarries were reopened and crushing plant was imported from Europe to speed the process. Railways were built to take the aggregate to the construction sites. A particular feature of this time was the excavation of several storage tunnels driven in the valley sides and totalling some 10 km in length; most of these are now sealed but one, in St Lawrence, was equipped as a hospital and is open to the public.

Since the war the quarrying industry has declined as fabricated materials have replaced natural stone in building and construction. A few working quarries remain and in recent years have supplied large amounts of rock for harbour works and land reclamation schemes at St Helier.

Siliceous flint-like rock, derived from French Oligocene

freshwater deposits, was imported in Medieval and later times to provide the famous French burrstones for the local mills.

Sand and gravel

Removal of sand and gravel from the Jersey beaches is forbidden by law, but there is no doubt that large amounts were taken in the past for building purposes. During the second world war sand was removed from the beach at Grouville for use in the construction of fortifications. Some of this sand was used locally, but much was taken by barge to St Helier and St Aubin, whence it went by lorry or by rail to construction sites in the west of the island (Ginns, 1973). More recently, some sand was taken from St Ouen's Bay and Grève de Lecq when the Val de la Mare dam [578 518] was constructed.

There are two major deposits of blown sand on the island, at St Ouen's Bay and at Grouville, and the greatest known thickness of 27 m was recorded at St Ouen's Bay (Keen, 1981; see also pp.92–93). Both deposits have been worked from time to time and some of the abandoned pits at St Ouen have been used for waste disposal. Concern about the uncontrolled exploitation of the sand at St Ouen led the States of Jersey to commission a report by the Institute of Geological Sciences on the resource potential of the area (Thurrell, 1972). The ecological significance of these dunes has been assessed by Ranwell (1975, 1976).

Brick clay and brickearth

Some 'clay' (a fine silt, probably redeposited loess) is present in the alluvial deposits in the St Helier basin and at St Ouen, and there is also some clay beneath shingle around the south and east shores of the island. These freshwater clays were formed in lagoons when sea level was lower than at present and in situations similar to those now obtaining at St Ouen's Pond. In addition the loess has provided brickearth of a quality suitable for making bricks; the most extensive unbroken sheet of loess – which is a silt rather than a clay – is in the district of Maufant (Old French meaning 'bad mud'), the name deriving 'from the nature of the soil ... which in flat areas tends to become waterlogged' (Mourant, 1935, p.493).

The clay in the St Helier basin was worked first, and the brickearth deposits were exploited from about the time of the Napoleonic wars, at Gallows Hill and West Mount, and most extensively in the Five Oaks area of St Saviour and at Mont à l'Abbé north of St Helier. When the loess at the brickworks at Five Oaks was exhausted, the underlying head and weathered andesite were hydraulically compressed into bricks.

Limestone

Eocene limestone occurs on the sea bed around Jersey, but limestone is absent from the island itself. Therefore limestone was formerly imported from France, particularly Carboniferous Limestone from Regnéville, to be burned locally to produce lime. Large quantities of local marine shells were also burned at one time for this purpose. As recently as 1918 mortar was made from a mixture of crushed and sifted granite and imported lime, burnt locally (Dr A. E. Mourant, personal communication). Jurassic limestones from Dorset were much used in military construction.

Fuel

Fossil fuels are virtually absent from Jersey, and wood was probably the staple fuel until coal and oil were imported. There are records of coal being shipped to Jersey and Guernsey from Neath and Swansea in Elizabethan times (Williams, 1934). During the second world war, and especially in 1944 – 45 when fuel imports were impossible, an attempt was made to extract the peat in St Ouen's Bay in order to supplement the wood supply, but the enterprise proved to be both impractical and expensive, each rising tide flooding the trenches that had been dug. The peat beneath Grouville Marsh was also investigated and eventually some hundreds of tons were dug, but the primitive method of extraction made the fuel very expensive. The principal fuels now imported are coal, oil and natural gas.

Minerals

A list of the minerals that occur in Jersey has been compiled over the years and an account of them was given by Mourant (1961, 1978). Although metallic minerals do occur, there has been little metalliferous mining in Jersey, and such attempts as have been made were unprofitable and short-lived. Ixer (1980) has described the occurrence of ore minerals in the island and has distinguished five types of mineralisation:

1   Pervasive copper mineralisation associated with rhyolitic and andesitic volcanic rocks, for example at Bouley Bay. The principal minerals are chalcopyrite, 'idaite' and minor enargite group minerals. Mineralogically and texturally these deposits resemble the porphyry copper type of mineralisation.

2   Molybdenite-pyrite-chalcopyrite-sphalerite mineralisation associated with granite pegmatite and occurring as
veins and in vugs within granite.
3   Vein mineralisation in granite, mainly magnetite-
quartz and minor sulphides. This mineralisation is later
than 2.
4   Minor sulphide-calcite mineralisation in the Jersey
Shale Formation.
5   Zinc-lead-silver-antimony mineralisation in the Jersey
Shale Formation at Le Pulec, genetically associated with
granite.

   Mourant and Warren (1934) described the history of mining in Jersey and the other Channel Islands. The best-known
and the most successful attempt at metalliferous mining in
Jersey was at Le Pulec [547 550], where three veins bearing
galena and sphalerite were discovered on the foreshore in
1871. Shingle was removed so that the veins could be examined by a mining engineer brought from Cornwall: good
silver values were reported by him (Williams, 1871) and by
Ogier (1871), and proposals were made to work the ore. One
plan to sink a shaft on land and extend from it a cross-cut to
reach the lodes beneath the bay was never implemented,
though the veins were worked on the beach at low tide.
Despite the fact that some ore was exported, the operation
seems to have been largely unsuccessful, for it was soon
abandoned (Mourant and Warren, 1934).
   Detailed descriptions of the Le Pulec mineralisation by
Ixer and Stanley (1980) and Stanley and Ixer (1982) show
that the sediments of the Jersey Shale Formation have
undergone two periods of polyphase mineralisation. In the
older, pyrite-pyrrhotine-marcasite were formed, and in the
younger arsenopyrite-chalcopyrite-sphalerite, along with
silicification and dolomitisation. The main ore-bearing veins
at Le Pulec post-date these earlier phases, and comprise
sphalerite and ferroan dolomite accompanied by polymetal-
lic mineralisation from fluids rich in Pb, Cu, Fe, Ag, and Sb.
This type of mineralisation is unique in Jersey.

Water supply

The water supply of Jersey comes mainly from the impounding of streams by dams constructed by the Jersey Water-
works Company. In St Helier formerly, and to some extent
at present, water has been derived from shallow wells (up to
about 10 m deep) tapping an aquifer in Pleistocene deposits
in the rock basin beneath the town. In about 1920, overpump-
ing of wells near the centre of the town drew in sea-water,
and wells near the coast became salty and were abandoned.

In the plateau areas, water was formerly obtained by impounding the headwaters of small streams. Now, outside the areas served by the water authority, increasing use is made of deep bored wells. The depth to the water table and the probable yield are difficult to predict; much seems to depend on whether a borehole intersects one of the irregularly distributed water-bearing fissures in the rocks. A little water is drawn from the aquifer formed by blown sand near St Ouen's Bay (see pp.92–93), and during droughts water is also obtained from the sea-water desalination plant at La Rosière [5578 4805].

References

ADAMS, C. J. D. 1967. A geochronological and related isotopic study of rocks from north-western France and the Channel Islands (United Kingdom). Unpublished D.Phil. thesis, University of Oxford.

— 1976. Geochronology of the Channel Islands and adjacent French mainland. *J. Geol. Soc. London*, Vol.132, 233–250.

ALLEN, P. 1972. Wealden detrital tourmaline: implications for northwestern Europe. *Q. J. Geol. Soc. London*, Vol.128, 273–294.

ANGUS, N. S. 1962. Ocellar hybrids from the Tyrone Igneous Series, Ireland. *Geol. Mag.*, Vol.99, 9–26.

ANSTED, D. T. and LATHAM, R. H. 1862. Geology of the Channel Islands. 247–297 in *The Channel Islands*. (London: W. H. Allen & Co.)

BACON, M. 1975. A gravity survey of the English Channel between Lyme Regis and St Brieuc Bay. *Philos. Trans. R. Soc. London*, A279, 69–78.

BARROIS, C. 1895. Le calcaire de Saint-Thurial (Ille-et-Vilaine). *An. Soc. Géol. N.*, Vol.23, 38–46.

BIRNIE, J. F., JONES, R. L., KEEN, D. H., and WATON, P. V. *In preparation*. Flandrian vegetational history and sea level change in Jersey, Channel Islands.

BISHOP, A. C. 1964a. The petrogenesis of hornblende-mica-lamprophyre dykes at South Hill, Jersey, C.I. *Geol. Mag.*, Vol.101, 302–313.

— 1964b. The La Collette sill, St Helier, Jersey, C.I. *Annu. Bull. Soc. Jersiaise*, Vol.18, 418–428.

— and KEY, C. H. 1983. Nature and origin of layering in the diorites of SE Jersey, Channel Islands. *J. Geol. Soc. London*, Vol.140, 921–937.

— and MOURANT, A. E. 1979. Discussion on the Rb-Sr whole rock age determination of the Jersey Andesite Formation, Jersey, C.I. *J. Geol. Soc. London*, Vol.136, 121–122.

— ROACH, R. A. and ADAMS, C. J. D. 1975. Precambrian rocks within the Hercynides. *In* A correlation of the Precambrian rocks of the British Isles. HARRIS, A. L. and others. *Spec. Rep. Geol. Soc. London*, No.6, 102–107.

BLAND, A. M. 1984. Field relationships within the South-west Jersey Granite Complex. *Proc. Ussher Soc.*, Vol.6, 54–59.

BLAND, B. H. 1984. *Arumberia* Glaessner & Walter, a review of its potential for correlation in the region of the Precambrian-Cambrian boundary. *Geol. Mag.*, Vol.121, Part 6, 625–633.

— EVANS, G., GOLDRING, R., MOURANT, A. E., RENOUF, J. T. and SQUIRE, A. D. 1987. Supposed Precambrian trace fossils from Jersey, Channel Islands. *Geol. Mag.*, Vol.124, 173.

BOUMA, A. H. 1962. *Sedimentology of some flysch deposits; a graphic approach to facies interpretation*. (Amsterdam: Elsevier.)

Briden, J. C., Clark, R. A. and Fairhead, J. D. 1982. Gravity and magnetic studies in the Channel Islands. *J. Geol. Soc. London*, Vol.139, 35–48.

Calvez, J.-Y. 1976. Comportement des systèmes uranium-plomb et rubidium-strontium dans les orthogneiss d'Icart et de Moëlan (Massif Armoricain). Thèse Docteur en Troisième cycle, Université de Rennes.

Casimir, M. 1934. Studies in folding of the Jersey Shales. *Proc. Geol. Assoc.*, Vol.45, 162–166.

— and Henson, F. A. 1955. The volcanic and associated rocks of Giffard Bay, Jersey, Channel Islands. *Proc. Geol. Assoc.*, Vol.60, 30–50.

Chappell, B. W. and White, A. J. R. 1974. Two contrasting granite types. *Pac. Geol.*, Vol.8, 173–174.

Cobbing, B. W. and Pitcher, W. S. 1972. The coastal batholith of central Peru. *J. Geol. Soc. London*, Vol.128, 421–460.

Cogné, J. 1959. Données nouvelles sur l'Antécambrien dans l'Ouest de la France. Pentévrien et Briovérien en baie de Saint-Brieuc (Côtes-du-Nord). *Bull. Soc. Géol. Fr.*, Vol.7, part 1: 112–118.

Coope, G. R., Jones, R. L. and Keen, D. H. 1980. The palaeoecology and age of peat at Fliquet Bay, Jersey, Channel Islands. *J. Biogeogr.*, Vol.7, 187–195.

— Jones, R. L., Keen, D. H. and Waton, P. V. 1985. The flora and fauna of late Pleistocene deposits in St Aubin's Bay, Jersey, Channel Islands. *Proc. Geol. Assoc.*, Vol.96, 315–321.

Cordell, L. 1970. Iterative three-dimensional solution of gravity anomaly data. USGS computer contribution program number W9303. Washington.

Cornes, H. W. 1933. The age and origin of the Jersey conglomerate. *Bull. Annu. Soc. Jersiaise*, Vol.12, 118–151.

Davies, K. H. 1983. Amino acid analysis of Pleistocene marine Mollusca from the Gower Peninsula. *Nature, London*, Vol.302, 137–139.

— and Keen, D. H. 1985. The age of Pleistocene marine deposits at Portland, Dorset. *Proc. Geol. Assoc.*, Vol.96, 217–225.

Day, A. A. 1955. On the values of gravity at St Anne (Alderney), St Peter Port (Guernsey), and St Helier (Jersey). *Mon. Not. R. Astron. Soc. Geophys. Suppl.*, Vol.7, 76–79.

— 1959. Gravity anomalies in the Channel Islands. *Geol. Mag.*, Vol.96, 89–98.

Dewey, J. F. 1969. Evolution of the Appalachian/Caledonian orogen. *Nature, London, Phys. Sci.*, Vol.22, 124–129.

Duff, B. A. 1978. Rb-Sr whole-rock age determination of the Jersey Andesite Formation, Jersey, C.I. *J. Geol. Soc. London*, Vol.135, 153–156.

— 1979. The palaeomagnetism of Cambro-Ordovician red beds, the Erquy Spilite Series and the Trégastel – Ploumanac'h granite complex, Armorican Massif (France and the Channel Islands). *Geophys. J. R. Astron. Soc.*, Vol.59, 345 – 365.

— 1980. The palaeomagnetism of Jersey volcanics and dykes, and the Lower Palaeozoic apparent polar wander path for Europe. *Geophys. J. R. Astron. Soc.*, Vol.60, 355 – 375.

— 1981. Scattered palaeomagnetic directions acquired during dioritization and stoping of the diorite-metagabbro complex, Jersey, C.I. *J. Geol. Soc. London*, Vol.138, 485 – 492.

DUMARESQ, P. 1935 (1685). A survey of ye Island of Jersey. *Bull. Annu. Soc. Jersiaise*, Vol.12, 415 – 416.

DUNLOP, A. 1911. On the Pleistocene beds of Jersey. *Bull. Annu. Soc. Jersiaise*, Vol.7, 112 – 120.

FURNIVAL, W. J. 1904. Jersey chinastone. 320 – 323 in *Leadless decorative tiles, faience and mosaic*. (Stone, Staffordshire.)

GINNS, M. 1973. Grouville Common during the German occupation. *Bull. Annu. Soc. Jersiaise*, Vol.21, 195 – 199.

GRAINDOR, M. J. 1957. Le Briovérien dans le nord-est du massif armoricain. *Mem. Expl. Carte Géol. Fr.* 211pp.

GROVES, A. W. 1927. The heavy minerals of the plutonic rocks of the Channel Islands. I (Jersey). *Geol. Mag.*, Vol.64, 241 – 251.

— 1930. The heavy mineral suites and correlation of the granites of northern Brittany, the Channel Islands and the Cotentin. *Geol. Mag.*, Vol.67, 218 – 240.

HAWKES, J. 1938. *The archaeology of the Channel Islands,* Vol.II, Jersey. (Jersey: Société Jersiaise.)

HELM, D. G. 1983. The structure and tectonic evolution of the Jersey Shale Formation, St Ouen's Bay, Jersey, Channel Islands. *Proc. Geol. Assoc.*, Vol.94, 201 – 216.

— 1984. The tectonic evolution of Jersey, Channel Islands. *Proc. Geol. Assoc.*, Vol.95, 1 – 15.

— and PICKERING, K. T. 1985. The Jersey Shale Formation—a late Precambrian deep-water siliciclastic system, Jersey, Channel Islands. *Sediment. Geol.*, Vol.43, 43 – 66.

HENSON, F. A. 1956. The geology of south-west Jersey, Channel Islands. *Proc. Geol. Assoc.*, Vol.67, 266 – 295.

INSTITUTE OF GEOLOGICAL SCIENCES. 1979. Bouguer gravity anomaly map, 1:250 000 Series (Revised edition), Guernsey Sheet 49N 04W.

IXER, R. A. 1980. The ore minerals of Jersey. *Annu. Bull. Soc. Jersiaise*, Vol.22, 443 – 451.

— and STANLEY, C. J. 1980. Mineralization at Le Pulec, Jersey, Channel Islands. *Mineral. Mag.*, Vol.43, 1025 – 1029.

KEEN, D. H. 1975. Two aspects of the last interglacial in Jersey. *Annu. Bull. Soc. Jersiaise*, Vol.21, 392 – 396.

— 1978a. A Palaeolithic flint flake from Noirmont Point. *Annu. Bull. Soc. Jersiaise*, Vol.22, 205 – 208.

— 1978b. The Pleistocene deposits of the Channel Islands. *Rep. Inst. Geol. Sci.*, No.78/26.

— 1981. The Holocene deposits of the Channel Islands. *Rep. Inst. Geol. Sci.*, No.81/10.

— 1982. Late Pleistocene land Mollusca in the Channel Islands. *J. Conchol.*, Vol.31, 57 – 61.

— HARMON, R. S. and ANDREWS, J. T. 1981. U-series and amino-acid dates from Jersey. *Nature, London*, Vol.289, 162 – 164.

KEY, C. H. 1974. The layered diorites of Jersey, Channel Islands. Unpublished PhD thesis, University of London.

— 1977. Origin of appinitic pockets in the diorites of Jersey, Channel Islands. *Mineral. Mag.*, Vol.41, 183 – 192.

LAUTRIDOU, J.-P. 1973. Les loess du Riss dans le bassin de la Seine. Proceedings of the IXth INQUA Congress, Christchurch, NZ, 54 – 55.

LAUTRIDOU, J.-P. (editor). 1982. The Quaternary of Normandy. Field handbook of the QRA meeting in Normandy, May 1982. (Cambridge: Quaternary Research Association.)

LEES, G. J. 1982. Mica-lamprophyres of south-western England. 301 – 302 in *Igneous rocks of the British Isles* SUTHERLAND, D. S. (editor). (Chichester: Wiley.)

LEFORT, J. P. 1975. Etude géologique du socle anté-mésozoïque au nord du massif Armoricain; limites et structures de la Domnonée. *Philos. Trans. R. Soc. Lond.*, A279, 123 – 135.

LEUTWEIN, F. 1968. Contribution à la connaissance du Précambrien récent en Europe Occidentale et développement géochronologique du Briovérien en Bretagne (France). *Can. J. Earth Sci.*, Vol.5, 673 – 682.

— and SONET, J. 1965. Contribution à la connaissance de l'évolution géochronologique de la partie nord-est du massif Armoricain français. *Sci. Terre, Nancy*, Vol.10, 345 – 367.

MASSON SMITH, D., HOWELL, P. M. and ABERNETHY-CLARK, A. B. D. E. 1974. The National Gravity Reference Network 1973. (Southampton: Ordnance Survey.)

MITCHELL, A. H. and READING, H. G. 1971. Evolution of island arcs. *J. Geol.*, Vol.79, 283 – 284.

MOURANT, A. E. 1932. The spherulitic rhyolites of Jersey. *Mineral. Mag.*, Vol.23, 227 – 238.

— 1933. The geology of eastern Jersey. *Q. J. Geol. Soc. London*, Vol.89, 273 – 307.

— 1935. The Pleistocene deposits of Jersey. *Annu. Bull. Soc. Jersiaise*, Vol.12, 489 – 496.

— 1940. *In* ROBINSON, A. J. Report of the Geological Section, 1939. *Bull. Annu. Soc. Jersiaise*, Vol.14, 13 – 15.

— 1956. The use of Ecréhous stone in Jersey. *Bull. Annu. Soc. Jersiaise*, Vol.16, 373 – 376.

— 1961. The minerals of Jersey. *Annu. Bull. Soc. Jersiaise*, Vol.18, 69 – 90.

— 1977. The use of Fort Regent granite in megalithic monuments in Jersey. *Annu. Bull. Soc. Jersiaise*, Vol.22, 41–49.

— 1978. *The minerals of Jersey.* (St Helier, Jersey: Société Jersiaise.)

— and WARREN, J. P. 1934. Minerals and mining in the Channel Islands. *Rep. Trans. Soc. Guernesiaise*, Vol.12 (for 1933), 73–88.

MUTTI, E. and RICCI LUCCHI, F. 1972. Le Torbitit dell'Apennino settenrionale; introduzione all'analisi di facies. *Mem. Soc. Geol. Ital.*, Vol.11, 161–199.

NORMARK, W. R. 1978. Fan valleys, channels and depositional lobes on modern submarine fans: characters for recognition of sandy turbidite environments. *Bull. Am. Assoc. Pet. Geol.*, Vol.62, 912–931.

NOURY, CH. 1886. *Géologie de Jersey.* (Paris: F. Savy; Jersey: Le Feuvre.)

OGIER, E. F. 1871. Rapport sur les mines de plomb du Pulec. (Jersey.) 1–8.

OLIVER, R. L. 1958. Andradite from the island of Jersey. *Annu. Bull. Soc. Jersiaise*, Vol.17, 181–184.

PICKERING, K. T. 1981. Two types of outer fan lobe sequence, from the late Precambrian Kongsfjord Formation submarine fan, Finnmark, North Norway. *J. Sediment. Petrol.*, Vol.51, 1277–1286.

— 1983. Transitional submarine fan deposits from the late Precambrian Kongsfjord Formation submarine fan, NE Finnmark. N Norway. *Sedimentology*, Vol.30, 181–199.

PLEES, W. 1817. *An account of the island of Jersey.* (Southampton.)

PLYMEN, G. H. 1921. The geology of Jersey. *Proc. Geol. Assoc.*, Vol.32, 151–172.

RANWELL, D. S. 1975. The dunes of St Ouen's Bay, Jersey: an ecological survey, part I. History and plant communities least modified by human influence. *Annu. Bull. Soc. Jersiaise*, Vol.21, 381–391.

— 1976. The dunes of St Ouen's Bay, Jersey: an ecological survey, part II. Plant communities strongly modified by man, the whole flora and management implications. *Annu. Bull. Soc. Jersiaise*, Vol.21, 505–516.

RENOUF, J. T. 1969. Geological report for 1968. *Annu. Bull. Soc. Jersiaise*, Vol.20, 15–16.

— 1974. The Proterozoic and Palaeozoic development of the Armorican and Cornubian provinces. *Proc. Ussher Soc.*, Vol.3, Part 1, 6–43.

— and BISHOP, A. C. 1971. The geology of the Fort Regent road tunnel. *Annu. Bull. Soc. Jersiaise*, Vol.20, 275–283.

RICHARDSON, K. 1984. The sedimentology and structure of the Rozel Conglomerate. Unpublished report, Goldsmiths' College, London.

ROACH, R. A., ADAMS, C. J. D., BROWN, M., POWER, G. and
RYAN, P. 1972. The Precambrian stratigraphy of the
Armorican massif, northwest France. *Proc. Int. Geol. Congr.,
Sess. 24, Montreal 1972*, 246–252.

ROBINSON, A. J. 1960. Geological report for 1959. *Annu. Bull.
Soc. Jersiaise*, Vol.17, 290–292.

ROSS, C. S. and SMITH, R. L. 1961. Ash flow tuffs—their
origin, geologic relations and identification. *U.S. Geol. Surv.
Paper* 366. 81 pp.

RYBOT, N. V. L. 1926. The corbels [of Grosnez Castle,
Jersey]. *Bull. Annu. Soc. Jersiaise*, Vol.10, 293–296.

— 1947. The quarrying and splitting of rocks in Jersey. *Bull.
Annu. Soc. Jersiaise*, Vol.14, 283–292.

SHACKLETON, N. J. and OPDYKE, N. D. 1973. Oxygen isotope
and palaeomagnetic stratigraphy of equatorial Pacific Core
V28–238: Oxygen isotope temperatures and ice volumes on a
10^5 and 10^6 year scale. *Quat. Res.*, Vol.3, 39–55.

SHOTTON, F. W. and WILLIAMS, R. E. G. 1971. Birmingham
University radiocarbon dates V. *Radiocarbon*, Vol.13, Part 2,
141–156.

SMITH, H. G. 1933. Some lamprophyres of the Channel
Islands. *Proc. Geol. Assoc.*, Vol.44, 121–130.

— 1936a. The South Hill lamprophyre, Jersey.
Geol. Mag., Vol.73, 87–91.

— 1936b. New lamprophyres and monchiquites from
Jersey. *Q. J. Geol. Soc. London*, Vol.92, 365–383.

SPEIGHT, J. M., SKELHORN, R. R., SLOAN, T. and KNAPP, R. J.
1982. The dyke swarms of Scotland. 449–459 in *Igneous rocks
of the British Isles*. SUTHERLAND, D. S. (editor). (Chichester:
Wiley).

SQUIRE, A. D. 1970. The sedimentology, provenance and age
of the Rozel Conglomerate, Jersey, Channel Islands.
Unpublished report, Chelsea College, London.

— 1973. Discovery of late Precambrian trace fossils in Jersey,
Channel Islands. *Geol. Mag.*, Vol.110, 223–226.

— 1974. Brioverian sedimentology and structure of Jersey and
adjacent areas. Unpublished PhD thesis, University of London.

STANLEY, C. J. and IXER, R. A. 1982. Mineralization at Le
Pulec, Jersey, Channel Islands; No.1 Lode. *Mineral. Mag.*,
Vol.46, 134–136.

STEVENS, J. 1965. *Old Jersey Houses*. Vol.I. (Chichester:
Phillimore.)

— 1977. *Old Jersey Houses*. Vol.II. (Chichester: Phillimore.)

TEILHARD DE CHARDIN, P. 1920. Sur la structure de l'île de
Jersey. *Bull. Soc. Géol. Fr.*, 4ᵉ Série, Vol.19 (for 1919),
273–278.

THOMAS, G. M. 1977. Volcanic rocks and their minor
intrusives, eastern Jersey, Channel Islands. Unpublished PhD
thesis, University of London.

THURRELL, R. G. 1972. The sand resources of St Ouen's Bay, Jersey. Unpublished report, Institute of Geological Sciences, London.

WALKER, R. G. 1978. Deep-water sandstone facies and ancient submarine fans: models for exploration for stratigraphic traps. *Bull. Am. Assoc. Pet. Geol.*, Vol.62, 932–966.

— 1984. In Facies models, Second Edition, *Geosci. Can. Reprint Ser., Geol. Assoc. Can.*, No.1, 317pp. WALKER, R. G. (editor).

WELLS, A. K. and BISHOP, A. C. 1955. An appinitic facies associated with certain granites in Jersey, Channel Islands. *Q. J. Geol. Soc. London*, Vol.111, 143–166.

WILLIAMS, B. 1871. *Report on the Jersey silver-lead mine.* (Jersey)

WILLIAMS, T. D. 1934. Trade relations between Jersey, Guernsey and Welsh ports in Elizabethan times. *Bull. Annu. Soc. Jersiaise*, Vol.12, 261–270.

ZEUNER, F. E. 1940. The age of Neanderthal Man, with notes on the Cotte de St Brelade, Jersey, C.I. *Occas. Pap. Inst. Archaeol.*, No.3.

Glossary

Acicular   Needle-shaped.

Adamellite   Granite in which alkali feldspar and plagioclase occur in about equal amounts.

Agglomerate   A volcanic rock formed of pyroclastic blocks or fragments generally more than 64 mm in diameter.

Air-fall tuff   A tuff formed by consolidation of fine-grained pyroclastic debris (ash) which was laid down on land from the air.

Amphibolite   A metamorphic rock consisting mainly of amphibole and plagioclase.

Amygdale   A gas bubble or cavity in an igneous rock which has been filled with secondary minerals.

Andesite   A lava of intermediate composition consisting of plagioclase feldspar (usually andesine) and one or more ferromagnesian minerals.

Andinotype   An orogeny characterised by sedimentation in fault-margined furrows, andesites, I-type tonalites, burial metamorphism, cauldron batholiths that feed volcanoes, and vertical movement with minimal shortening.

Anticline   A fold, generally convex upward (an arch), the core of which contains the older rocks.

Aphyric   Adjective applied usually to a fine-grained igneous rock which lacks phenocrysts.

Aplite   A light-coloured, fine- to medium-grained acid igneous rock with an equigranular texture and consisting mainly of quartz and feldspar.

Aplogranite   A light-coloured, even-grained plutonic acid igneous rock consisting mainly of quartz and alkali feldspar.

Apophysis   An offshoot from an igneous intrusion.

Appinite   A dark-coloured plutonic igneous rock rich in hornblende which commonly occurs as elongate prismatic crystals.

Ash   Unconsolidated fine-grained pyroclastic debris.

Assimilation   Incorporation of foreign material into a magma.

Aureole   The rocks adjoining an igneous intrusion that have suffered contact metamorphism.

Authigenesis   The process whereby minerals are formed in place within a sedimentary rock during or after deposition.

Autobreccia   A rock composed of angular fragments, formed by a process that is penecontemporaneous with the deposition or consolidation of the rock.

Axial plane   The surface that passes through successive hinge lines within a fold.

Axial trace   The line of intersection of the axial plane or axial surface of a fold with the Earth's surface.

Axiolite   A spherulitic aggregate elongated along a central axis.

Basalt   A fine-grained lava or minor intrusion composed mainly of calcic plagioclase and pyroxene with or without olivine.

Base-surge   A cloud of gas and solid debris that moves rapidly outward from the base of a volcanic explosion column.

Bouguer anomaly   A gravity anomaly left after corrections have been made for the attraction effect of topography.

Braid   To branch and rejoin repeatedly, forming a network.

Breccio-conglomerate   A sedimentary rock consisting of angular and rounded sedimentary clasts.

Camptonite   A lamprophyre consisting mainly of plagioclase (usually labradorite) and brown hornblende, the hornblende forming elongated prismatic crystals.

Cataclasite   A rock formed by shattering less severe than would produce a mylonite.

China-stone   Any form of granitic rock used in the manufacture of china and usually, but not necessarily, kaolinised.

Clast   A grain or fragment in a sedimentary rock.

Cleavage   Aligned and closely spaced tectonic surfaces along which a rock tends to split.

Columnar jointing   Prismatic fractures in lavas, sills or dykes that have resulted from cooling.

Concretion   A nodular or irregular mass formed by the secondary precipitation of minerals about a nucleus or centre in a sedimentary rock.

Conglomerate A sedimentary rock consisting of cemented rounded pebbles or clasts.

Contact metamorphism Metamorphism resulting from the emplacement of a body of magma.

Cross-cut A mine tunnel driven through barren rock, commonly to intersect a mineral deposit.

Crystal fractionation The separation of a cooling magma into parts of differing composition by the successive crystallisation and settling of different minerals at progressively lower temperatures.

Devitrification The replacement of glassy texture by crystalline texture in a volcanic rock during or after cooling.

Dextral fault A fault in which the rock on the far side appears to have been moved horizontally to the right.

Diagenesis The sum of the processes involved in changing a sediment into a sedimentary rock.

Diorite A plutonic igneous rock of intermediate composition consisting essentially of plagioclase and hornblende, with or without biotite and pyroxene.

Disconformity A break in the stratigraphic sequence without major structural discordance.

Dolerite A medium-grained igneous rock consisting mainly of calcic plagioclase and pyroxene, commonly with an ophitic texture, with or without olivine.

Dolmen A prehistoric burial chamber.

Drag fold A fold produced by differential movement between beds on the limb of a large fold or by shearing in a fault zone.

Dune A ripple-like bedform greater than 10 cm in height and 1 m in wavelength.

Enclave An inclusion or fragment enclosed in an igneous rock.

Epiclastic An adjective applied to sedimentary rock formed of fragments derived by weathering and erosion of older rocks.

Epidiorite A metamorphosed basic igneous rock, consisting principally of amphibole and plagioclase.

Eugeosynclinal A major elongate structural and sedimentological unit of the Earth's crust, with a thick sequence of deep-water sediments and characteristic igneous rocks.

Euhedral An adjective applied to mineral grains in igneous rocks that are bounded by their natural crystal faces.

Eutaxitic texture The texture in tuffs where shards and pumice are flattened to give a banded or streaky appearance.

Facies The lithology, structure, fauna, etc., of a rock unit.

Felsite A fine-grained igneous rock composed mainly of quartz and feldspar.

Fiamme Collapsed pumice fragments in ignimbrite, commonly with ragged terminations.

Flow banding A structure characterised by alternating layers of slightly different composition and texture owing to the movement of magma, most common in silicic lava flows.

Fluxion banding Flow banding.

Flysch A marine sedimentary facies comprising a thick sequence of sandstones, shales and mudstones, typically found on the borders of the Alps.

Fold axis or hinge-line The line at which the two sides or limbs of a fold meet.

Gabbro A coarse-grained intrusive igneous rock composed essentially of basic plagioclase and pyroxene with or without olivine.

Gneiss A foliated metamorphic rock in which layers of coarsely granular minerals alternate with layers or lenticles of platy minerals.

Graded bed A sedimentary unit in which the grains normally show a progression from coarse below to fine above. Some beds display inverse grading.

Granite In common usage, a coarse-grained acid igneous rock consisting essentially of quartz, feldspar and mica. More precisely, the feldspar is predominantly (greater than 2/3) alkali feldspar.

Granoblastic A textural term used of regionally metamorphosed rocks having mineral grains of the same general size.

Granophyre A granitic rock in which the bulk consists of micrographic intergrowths of quartz and potassic feldspar, giving a texture called granophyric.

Gravity gradient A measure of the change in the value of gravity relative to horizontal distance.

Greenschist facies Regionally metamorphosed rocks produced under conditions of low temperature and low to medium pressure.

Greywacke A poorly sorted sandstone with more than 15 per cent interstitial matrix and angular to subangular grains of quartz, feldspar and lithic fragments.

Heavy mineral A mineral with specific gravity greater than 2.9; a mineral that will sink in bromoform.

Hornfels A rock consisting of fine equidimensional grains without preferred orientation, produced by contact metamorphism.

Humic Derived from plants.

Hydrothermal Relating to hot solutions emanating from a magma.

Idiomorphic *See* euhedral.

Ignimbrite A subaerial pyroclastic rock formed by the deposition and consolidation of ash flows and glowing avalanches (nuées ardentes).

Imbrication A sloping and overlapping arrangement.

Intraclast A sedimentary rock fragment derived penecontemporaneously from within the sedimentary basin, e.g. pebbles in an intraformational conglomerate.

Isochron A straight line constructed most commonly by plotting the ratio $^{87}Rb/^{86}Sr$ against the ratio $^{87}Sr/^{86}Sr$, or $^{207}Pb/^{204}Pb$ or $^{208}Pb/^{204}Pb$ against $^{206}Pb/^{204}Pb$, for different rocks or minerals from the same geological body. The slope of the isochron is a function of the age of the body.

Isocline A fold with parallel limbs.

Isotopic dating The determination of the age of a rock by methods based on the nuclear decay of certain natural chemical elements contained within it.

Kaersutite A titanium-bearing amphibole.

Keratophyre A soda-rich acid to intermediate lava or minor intrusion.

Kersantite A lamprophyre consisting mainly of biotite and plagioclase, usually accompanied by augite and/or hornblende.

Lacustrine Relating to a lake.

Lahar A mudflow composed of volcaniclastic material.

Lamprophyre A general term for dark, porphyritic intrusive igneous rocks composed of phenocrysts of one or more of dark mica, pyroxene, amphibole or olivine in a groundmass of alkali feldspar or plagioclase. Some lamprophyres are feldspar-free.

Lapilli Fragments in the range of 5 to 50 mm ejected by volcanic eruption.

Liquefaction Quicksand effect in soft sediments owing to sudden increase in pore fluid pressure and loss of cohesion.

Lithic Made of rock.

Machair Low-lying sandy environment.

Mafic Rich in ferromagnesian minerals.

Magnetic anomaly A departure from the normal magnetic field of the Earth.

Megacryst A crystal significantly larger than the grains in the surrounding groundmass of an igneous rock.

Metasediment Metamorphosed sedimentary rock.

Metasomatism Metamorphism involving a change in the bulk composition of the affected rock.

Microlites, microliths Microscopic needle-like crystals generally found in volcanic rocks and having some determinable optical properties.

Minette A lamprophyre in which biotite forms phenocrysts and orthoclase is the main feldspar.

Molasse A sequence of sedimentary rocks laid down in intermontane basins.

Monchiquite A lamprophyre composed essentially of phenocrysts of pyroxene, olivine and titanium-bearing amphibole in an isotropic groundmass consisting largely of analcime, commonly highly altered.

Monzonite A coarse-grained igneous rock intermediate in composition between syenite and diorite, containing approximately equal amounts of potassic feldspar and plagioclase.

Mortar structure A structure produced by the mechanical fracture of rocks, especially granites and gneisses, such that the comminuted grains of quartz and feldspar surround larger grains of the same minerals.

Mylonite   A compact fine-grained streaked and flinty rock formed by severe granulation and shearing of rocks during dynamic metamorphism.

Neomagmatic   An adjective referring to magma formed by partial or complete refusion of pre-existing rocks.

Orthogneiss   A coarse-grained rock produced by the regional metamorphism of igneous rocks.

Palaeocurrent   A current that flowed in the geological past.

Palaeomagnetism   The study of natural remanent magnetisation in order to determine the intensity and direction of the Earth's magnetic field in the geological past.

Palaeoslope   A slope that existed in the geological past.

Parasitic fold   A relatively small fold on the limb or in the hinge of a larger congruous fold of the same generation.

Parataxitic texture   An extreme variation of eutaxitic texture in tuffs, in which the shards and pumice lumps are flattened and appear to have been drawn out.

Pegmatite   An exceptionally coarse-grained igneous rock. Most pegmatites are granitic and form irregular dykes, lenses or veins, especially near margins of intrusions.

Pelitic Argillaceous

Pericline (doubly plunging fold)   A fold in which the beds dip outwards from the centre (dome) or towards the centre (basin).

Perlitic texture   Small-scale arcuate cracks caused by cooling in volcanic glass.

Petrographical   Pertaining to the descriptive aspects of the study of rocks (petrology).

Phenocryst   A crystal in an igneous rock that is conspicuously larger than those of the matrix in which it is set.

Plunge   The inclination of a fold axis.

Plutonic   Relating to igneous rocks formed at great depth in the Earth.

Pneumatolysis   Alteration of a rock, or crystallisation of minerals, by gases emanating from a magma.

Poikilitic   A texture in which smaller crystals of one mineral are enclosed within a larger crystal of another.

Poikiloblastic   A texture in a metamorphic rock formed where a recrystallised mineral surrounds relicts of earlier minerals.

Polygenetic   Originating in more than one way.

Porphyritic   A texture in which larger crystals in an igneous rock are set in a finer-grained groundmass, as in a porphyry.

Porphyroblast   A large, usually well shaped crystal, that has grown in a finer-grained matrix during metamorphism.

Pressure solution.   The mass transfer of material by fluid diffusion from one part of a rock to another as a result of grain-to-grain contact during tectonic deformation.

Propylite   A hydrothermally altered andesite or related rock containing secondary minerals such as chlorite, zoisite and calcite.

Provenance   Source areas from which fragments in sedimentary rocks have been derived.

Proximal   Close to the source of supply of sedimentary material.

Pseudomorph   A mineral whose outward crystal form is that of another, which it has replaced.

Pumice   A highly vesiculated glassy lava, usually of rhyolite and light enough to float in water.

Pyroclastic   Adjective describing a clastic rock formed by explosion or eruption from a volcanic vent.

Quartz-wacke   A sedimentary rock with 15 per cent or more matrix in which framework grains are mainly quartz.

Radiocarbon date   The age of a deposit determined by measuring the content of carbon-14 (^{14}C) in organic material.

Raised beach   A beach deposit left at a level above that of the modern beach, following a lowering of sea level.

Rhyolite   A fine-grained acid extrusive igneous rock, commonly porphyritic and flow-banded.

Ripple mark   A small sand ridge formed by the movement of water or sediment over unconsolidated sediment. Ripples usually have a cross-laminated internal

ructure useful in determining the
alaeocurrent direction.

chist A strongly foliated metamorphic
ock in which the lamellar or elongate
ninerals show parallel orientation.

cree An accumulation of angular rock
ragments, usually at the foot of and
erived from a cliff or hill.

hard A glass fragment typically found
n pyroclastic rocks, having distinctive
uspate margins.

hear A fracture caused by shearing
ue to compressive stress.

inistral fault A fault in which the rock
n the far side appears to have been
noved horizontally to the left.

ole mark A sedimentary structure at
ne base of a bed produced by currents
nd used to determine the direction of
alaeocurrent flow.

pessartite A lamprophyre consisting
nainly of phenocrysts of hornblende set
n a groundmass of sodic plagioclase.

pherulite A mass of radiating crystals,
sually spherical in shape.

talagmite A column of calcareous
naterial deposited on the floor of a cave
y water dripping from the roof.

trike slip The component of the
novement on a fault that is parallel to the
trike of the fault.

ubhedral crystal A mineral within an
gneous rock only partly bounded by its
atural crystal faces.

ub-volcanic Relating to the region
elow the Earth's surface where dykes
nd sills are intruded.

upracrustal Descriptive of rock that
verlies the basement.

yenite A plutonic rock consisting
ssentially of alkali-feldspars and
mphibole or dark mica.

yncline A fold, generally convex
ownward, the core of which contains the
ounger rocks.

ynclinorium A composite regional
yncline composed of lesser folds.

ynplutonic A term used to describe a
yke intruded into a granite rock before
ne granite had solidified.

Tear fault A fault in which the
novement has been substantially in a
orizontal sense.

Tectonic Relating to the forces involved
in the large-scale structural evolution of
the upper part of the Earth's crust.

Tectonic pitting The indentation of
one sedimentary grain by another.

Trace fossil A sedimentary structure,
such as a track or boring, left by an
animal in the geological past.

Tuff A lithified deposit of volcanic ash.

Tuffite An admixture of pyroclastic
(> 25 per cent) and epiclastic (> 25 per
cent) material.

Turbidity current A turbulent current
laden with suspended sediment at the
base of a column of water that flowed
down a slope under the influence of
gravity.

Unconformity A break in the
stratigraphical sequence marked by a
structural discordance.

Vergence The direction in which a fold
is inclined or overturned.

Volcaniclastic Composed mainly of
volcanic rock fragments.

Wrench fault A tear fault.

Xenocryst A crystal in an igneous rock
to which it is not genetically related.

Xenolith An inclusion in an igneous
rock to which it is not genetically related.

Index of geographical localities

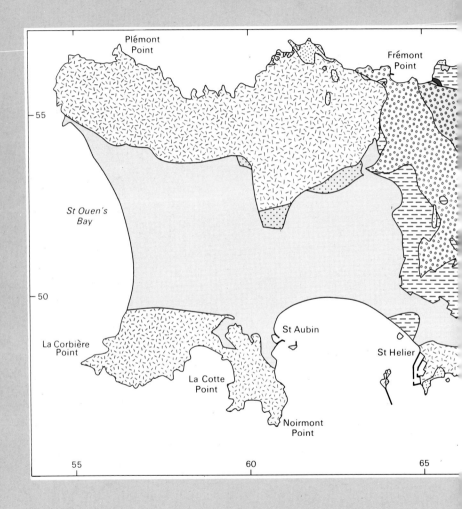

Plémont
Point

Frémont
Point

— 55

St Ouen's
Bay

— 50

La Corbière
Point

St Aubin

St Helier

La Cotte
Point

Noirmont
Point

55

60

65